ABOUT THE AUTHOR

After graduating from Cambridge and the University of Sussex, where he gained a doctorate in industrial economics at the university's Science Policy Research Unit, Tony Golding spent several years with an international electronics company. He then entered the City in a research role and spent 24 years there, largely with a London-based investment bank. He left in 1998 to write *The City: Inside the Great Expectation Machine*, which, in September 2002, went into a second edition. An early investment in Cambridge Mechatronics Limited, taking him back to his roots as a technology analyst and investor, inspired him to write *Down to the Wire*.

'Tony Golding has unearthed a little-known technology success and tells its story with great care and precision. It's a great reminder – if one were needed – that British ingenuity continues to play a vital role on the world stage.'

James Ashton, Author of *The Everything Blueprint: The Microchip Design that Changed the World*

'Like all the best business histories, this is far more than just the compelling story of one plucky British startup's struggle to commercialise its cutting-edge innovations. The challenges it faces sharply illuminate the profound geopolitical and technological shifts reshaping our world.'

Simon Nixon, Publisher of the Wealth of Nations newsletter, and former Chief Leader Writer at *The Times*

'Centred on a British company that manufactures a key component for smartphones, *Down to the Wire* clearly sets out the remarkable story of how and where handsets are made. It also reveals how China's rise as a major innovator presents both challenges and real opportunities for UK companies with global ambitions.'

Kerry Brown, Professor of Chinese Studies and Director of the Lau China Institute at King's College, London

'*Down to the Wire* is a fascinating study that unpacks the complexities of one of the 21st century's most transformative inventions. Tony Golding masterfully explains the rise of the smartphone and the challenges facing an innovative UK-based startup as it attempts to pivot towards emerging technology leaders in China.'

Professor Jeremy Hall, Director of the Science Policy Research Unit (SPRU), University of Sussex Business School

Down to the Wire

How a UK startup
hitched a ride
with China's tech
champion

Tony Golding

First published in 2026 by
Tony Golding, in partnership with Whitefox Publishing Limited

www.downtothewirecont.com

www.wearewhitefox.com

EU GPSR Authorised Representative
LOGOS EUROPE, 9 rue Nicolas Poussin, 17000, LA ROCHELLE, France
E-mail: Contact@logoseurope.eu

ISBN 978-1-9181916-3-9
Also available as an eBook
ISBN 978-1-9181916-4-6

Designed and typeset by Seagull Design
Cover design by Laura Thomas
Project management by Whitefox Publishing Limited

For Gill, Nicola, Andrew, Chloe and Rebecca

CONTENTS

INTRODUCTION

A CHINESE RE-AWAKENING

At the end of August 2023, the smartphone world was shaken by a wholly unexpected event. Huawei, the premier Chinese technology company and de facto national tech champion, announced its Mate 60 phone. Fuelled by online indications of a technological breakthrough, the Chinese rumour mill went into overdrive. The phone appeared to have features that were beyond Huawei's manufacturing capability – as a direct consequence of the technological embargo imposed by the USA on Huawei's access to the critical silicon chip inside a smartphone.

For one small British company, however, Huawei's smartphone breakthrough – as it turned out to be – was a cause for celebration. Cambridge Mechatronics Limited (CML), a UK startup, had invented and commercialised an innovative component used in smartphone cameras. After several years spent trying to persuade leading handset manufacturers to adopt CML's unique technology, the company had achieved its own breakthrough with Huawei (then the third largest phone maker globally) in 2018. It was initially designed into Huawei's flagship (top-of-the range) P30 phone launched in March 2019 and had since then become a mainstay of the company's flagship line-up.

For an 80-person UK company located 6,000 miles from Huawei's headquarters in Shenzhen, just north of Hong Kong, to have convinced Huawei to grant it a permanent slot on Huawei's flagship smartphones was – after more than two years of rigorous testing and evaluation – a remarkable achievement. At that stage, Huawei, despite China's growing technical capability – and the Chinese government's marked preference for locally sourced inputs – relied for certain functions on foreign components. But these were, in very large part, supplied by substantial, well-endowed, well-established firms – not a miniscule operation based in Cambridge.

This book is the story of Cambridge Mechatronics: its origin, how it invented a new type of component for smartphones and its struggle to persuade handset manufacturers to adopt it. CML is a classic entrepreneurial startup that evolved into a substantial small company. For directors, employees and shareholders, it has been a roller coaster ride, with plenty of blind alleys and false dawns. The company's saga can only be understood in the context of the industry it serves, so this book is as much about the smartphone – its evolution, the technology underlying it, the business strategies of the participants and where it might be going. You, the reader, will accompany CML on its technological and commercial journey as it tries to pick its way through the smartphone jungle. Sadly, the industry has become enmeshed in the contest for technological leadership between the United States and China. As a consequence, geopolitics casts a long shadow over CML's progress at several points in the narrative.

The timing of Huawei's announcement in August 2023 caught everyone by surprise as Huawei had previously followed a strict pattern of announcing the new Mate generation in September. Unusually, the new model was made available to the market immediately after its unveiling on the Huawei website rather than being released after a glitzy launch event. Information about its specifications, especially its communications capabilities and the central silicon chip it relied on, was sparse. But limited spec data did nothing to restrain Huawei enthusiasts. In an outpouring of patriotic fervour, all the available Mate 60s were swept off the shelves in the first few days.

Huawei's new phone was not supposed to have happened. For three years there had been a widespread industry conviction that it did not possess the necessary technology. Observers registered surprise, while in Washington DC there was consternation – and concern. In the space of less than a decade, China had secured pole position in smartphones. In 2019 the new smartphone order began to be buffeted by the winds of geopolitics as political and commercial differences between the United States and China deepened. Huawei, just a few years earlier one of the giants of smartphone manufacture, ranking alongside Samsung and Apple, found itself in the eye of the storm. These events changed the shape of the global smartphone industry and were the precursor to the *sotto voce* introduction of the Mate 60. Caught up in these happenings, Apple has engaged in a tricky but effective geopolitical high-wire act – until the business model it had fashioned was upended by President Trump's April 2025 tariff onslaught.

THE SMARTPHONE JUNGLE

This book is substantially about your smartphone, the clamorous metal, glass and plastic oblong that occupies a pivotal role in our twenty-first-century lives. It explains in a non-technical fashion what happens inside a handset to enable it to perform the amazing variety of functions we demand of it. It will appeal to anyone who has ever wondered about the little device they carry around with them and look at many times a day – how it works and how it is made.

Of course, calling the slab-like object we rely on a 'phone' is an anachronism. It has decisively cast off the shackles of its former life as a pure speaking instrument. The device that is constantly by our side is, in essence, a personal – or portable – communications and information device. Unfortunately, 'PCID' doesn't exactly roll off the tongue. 'Phone' is what we are stuck with. In 2025 an astonishing five billion people on the planet owned one, up from half that number in 2016.

The years since the original iPhone – the first mass-market smartphone – in 2007 have seen astounding technical progress. The communication and computing power we now hold in our hands is a modern miracle. As well as the technology embodied in a smartphone, the greater part of this book is about the *business* of developing, making and selling handsets. Since the first iPhone, the business of making smartphones has decisively shifted its centre of gravity, away from Europe, Japan and North America to China and South Korea. Nokia downsized its plants in Finland, Hungary and Mexico in 2012, and Motorola closed its Texas assembly plant in 2014.

After 2015, the emergence of Oppo, Vivo, Xiaomi, Huawei and others as vigorous home-grown competitors in smartphones, together with enhanced investment in its supply chain by Apple, entrenched China as the epicentre of the global smartphone manufacturing industry. Put bluntly, China made a land grab for the smartphone business – and succeeded. Europeans and Americans, familiar with Samsung's and Apple's leadership in their domestic markets, are at risk of thinking that Chinese phones occupy a secondary role globally. Far from it. Aside from the advanced nations, Chinese handsets dominate everywhere.

A top-of-the-range phone is an assemblage of around 1,500 individual components. Consequently, the manufacture of smartphones is a highly complex international collaboration. Hundreds of components and sub-systems from around the world are designed, produced and delivered on time in the right quantities, at the required level of quality, to assembly locations in Asia. The industry is characterised by a high degree of global interdependence. It is not an exaggeration to describe the smartphone supply chain as labyrinthine. America and the West generally are dependent on China and other countries in Asia but, as we shall see, several Chinese handset producers are similarly reliant on US-generated inputs – both components and technology.

MODERN ELECTRONICS – CHIPS AND ACTUATORS

Anyone who takes the slightest interest in modern electronics will appreciate that exceptional progress in semiconductors over the last 50 years lies at the core of the world we now live in and, in particular, is responsible for the smartphone's formidable capabilities. An integrated circuit (or IC) is the generic term for any semiconductor chip incorporating thousands, or nowadays billions, of transistors (tiny devices that act as a switch or an amplifier). In 2013, the iPhone 5's main semiconductor contained one billion transistors. The Apple A19 Pro chip, which powers the iPhone 17, has in excess of 20 billion transistors.

The key integrated circuit in your phone is called a System-on-a-Chip (invariably abbreviated to SoC). It functions as the 'brain' of the device. Few smartphone owners realise the debt they owe to a cluster of advanced semiconductor production facilities located in science parks strung along the north-west coast of Taiwan. Silicon Island – as it is justly dubbed by some – is responsible for the manufacture of 80 per cent of all smartphone SoCs, and a still higher proportion of the SoCs inside phones sold in Western markets. Most Chinese smartphone manufacturers are equally dependent on Taiwan for their SoCs. Chapter 3 provides an easy introduction to the world of semiconductors – the defining technology of our age.

The SoC is not the only chip inside your phone. A handset contains additional pieces of silicon, especially memory and wireless chips, backing it up. Most mobile users are unaware that the smartphone is also the product of numerous parallel innovations in a host of less glamorous areas that are essential to its functioning: displays, sensors, cameras, batteries, glass and mechanical components. To the consumer, these are invisible – and unsung. CML's component is one of these, occupying a position at the base of the supply chain pyramid.

The product invented by the engineering team at Cambridge Mechatronics is not purely electronic in nature but electro-mechanical, a component that translates electronic signals into physical movement. It is known as an actuator (fully explained in Chapter 1). Actuators are ubiquitous in our daily lives, operating invisibly in the background. Much innovation in the tech field nowadays takes the form of software, often an app. In contrast, CML's unique device is software-driven – as indeed are many electronic components – but the hardware is manifestly physical and has to be assembled on a production line. Getting the product right can be a painstaking process, with constant adjustments, improvements and iterations. It is not merely a matter of rewriting a few lines of code.

Central to CML's actuator – marking it out from existing actuators – is its reliance on wire. Not just any old wire but a hyper-intelligent piece of wire thinner than a human hair, controlled by a proprietary microchip to a high level of precision. In a very real sense, everything the company has achieved is (echoing the title of this book) 'down to the wire'. On its own, an actuator is inert. It requires an accompanying 'driver' semiconductor chip to feed it the precise electrical instructions it needs to perform in the application for which it has been designed. Creating and refining both products requires a multi-disciplinary skillset. Accordingly, the CML research team's expertise covers mechanical, electronic and control engineering together with software and materials science.

The new type of actuator that the team at Cambridge Mechatronics invented goes into smartphone cameras. It is used to move the lens

– a vital function in the eternal quest to match the functionality of traditional, stand-alone digital cameras. Consequently, to illustrate how the myriad components in a handset come together to create the device you hold in your hand, the focus in this book – pun fully intended – is on the camera.

For many consumers, the stand-out advance over the last 20 years is the mobile phone's greatly enhanced picture-taking ability. With the rise of the smartphone, amateur photographers ditched their cameras in droves, with dire consequences for the camera industry. During the 2010s digital camera sales were decimated. Annual sales are now less than a tenth of their peak level in 2010. According to a website called Photutorial, smartphones accounted for 94 per cent of the 1.94 trillion photos taken worldwide in 2024. (Quite how such a precise figure is calculated is not disclosed.) For smartphone photography, the year 2017 marked a major turning point. Propelled by a spate of technological advances, the single rear camera was joined by multiple sister cameras, opening up a world of opportunity. It was as if a single flower was suddenly surrounded by several new buds. Today, every handset has multiple cameras, so the market for CML's component is large.

CAMBRIDGE MECHATRONICS AND ME

More than 20 years ago I invested in Cambridge Mechatronics, then a company with 30 employees. The company's progress since then, has been – to put it mildly – uneven. *Down to the Wire* is deliberately evocative. It will resonate with shareholders because the company has constantly been on the verge of achieving its aims – and constantly on the edge of being able to raise funds to finance the next stage of its growth.

CML believes it has invented a unique technology with applications beyond smartphones that will win through in the end.

Fortunately, finance has been available to support the dream, originally the brainchild of a physicist who decided to pursue his vision. Until recently, the company relied almost entirely on private investors to fund its development. Credit is due to all those who invested in CML for their tenacity in the face of setbacks, but the commitment of one individual as the anchor investor in CML from 2004 stands out above all others. None of this talk about money, however, should obscure the simple truth that the achievements of CML are down to the skill demonstrated by management in steering the company through choppy waters and the cohesiveness of its original talented and dedicated core engineering team.

My status as a minor but engaged shareholder has allowed me to observe CML at close quarters for over two decades. Not from the inside, I emphasise, but as an investor who wanted to be kept in touch. I liken my vantage point to a spectator at a football match, albeit one with a seat in the front row of the stand. It has long been my feeling this was a tale worth telling, both as a record of an ambitious UK technology-based venture and because – reflecting my background as an industrial economist – corporate histories are intrinsically interesting.

Many corporate histories tend to be celebratory, glossing over the setbacks. While not a blow-by-blow account – that would be a lengthy and tedious read – this narrative does not shy away from recording the (numerous) disappointments amid the triumphs. For those engaged in business, there are undoubtedly lessons to be learnt from CML's experience. As such, it is my hope that the account will appeal to anyone interested in the raw commercial realities of running a small, ambitious, high-tech enterprise. I confine myself to a few observations at the end. You are welcome to draw your own conclusions.

The book focuses on CML's technological journey and the practical issues the company encountered, with considerably less emphasis on the personalities involved and the details of financing. It is not, in any sense, an official history. Several people associated with the company have read and commented on the text of this book, but it reflects my interpretation of events. Any views expressed are my own.

FROM B2C TO B2B

In the field of innovation there exists a fundamental distinction between business- and consumer-facing inventions. Selling a new product to a corporate entity is a wholly different proposition from trying to persuade consumers to take up something new. In the dot-com boom at the end of the 1990s, the terms Business-to-Consumer (B2C) and Business-to-Business (B2B) entered common currency. CML's original product was readily understandable by anyone with the slightest interest in audio: it offered consumer audio companies a superior way of generating surround sound. By the mid-2000s, however, the company had shifted its centre of gravity to focus on an entirely different product for a wholly different market: actuators for mobile phone cameras. In essence, CML, a decade after its founding, transitioned its business model from B2C to B2B.

For shareholders, CML's strategy shift represented a quantum leap in terms of comprehensibility, both in terms of the product itself and its target industry. When the company performed its pivot, I resolved to try, as best I could, to keep up with smartphone camera technology. Not so much from a technical standpoint but from the point of view of someone with an academic background in the economics of science and technology eager to understand the business implications of the relentless pace of innovation in electronics. In doing so, I have no doubt that much of what I have written about the technology of actuators, semiconductors and phones will be regarded, at best, as oversimplification and, at worst, plain wrong, despite my attempts to check with those who know.

The technology created and commercialised by Cambridge Mechatronics has been proven to work – reliably and effectively – in over 90mn devices. But as investors in the company know only too well, the road to financial viability has been strewn with obstacles. Looming over all these events has been the dead hand of geopolitics, which has weighed heavily on the company's fortunes. CML has yet to realise its potential but, as we shall see, it has now attracted high-grade, heavyweight backing, has moved into pastures new and hopes are running high.

SOURCES: SMARTPHONES IN CYBERSPACE

In writing this book I have relied heavily on the internet for source material. Cyberspace is host to a whirlpool of fact, analysis, comment, gossip, rumour and outright speculation on all aspects of smartphones posted by the massed legions of what I call the 'techno-commentariat'. Some of the more than a hundred websites comprising the techno-commentariat have names that describe broadly what they do (TechCrunch, TechRadar, Android Authority) while others are more self-consciously geeky (Gizmodo, Ubergizmo, ZDNET). New information is picked up and instantly relayed around the internet. The techno-commentariat is a rich resource but it is necessary to be cognisant of any agenda, commercial or otherwise, when evaluating the information.

Every Android smartphone maker – to a greater or lesser extent – trails details of forthcoming handsets to the techno-commentariat in the belief that this is an effective way of creating a buzz ahead of the official launch. Some even indulge in self-leaking – releasing information on upcoming models on websites with which they are associated. Apple's founder, Steve Jobs, by contrast, famously delighted in the Big Reveal. Consequently, Apple, consistent with the company's pervasive culture of secrecy, strives to prevent premature disclosure. With limited success. Apple goes to extreme lengths to prevent employees revealing confidential information about the company or its products. But it is far less able to police its 200-company strong iPhone component supply chain, which is notoriously leaky.

Smartphones and semiconductors are technical in nature and abbreviations are widely used for industry-specific terms. This book sometimes uses these for ease of reading. Appendix 1 lists the abbreviations most commonly encountered. Cambridge Mechatronics is referred to by its original name, 1...Limited, until the name was changed in 2008. After that, CML is used.

PART 1
AUDIO TIME

1.

In the beginning ...

TOWARDS TRANSDUCERS

What exactly is an actuator? It is a type of transducer. There is much electronics cannot achieve in the real world without being able to effect and control mechanical movement. Devices that produce movement are called mechanical transducers, or just transducers: components that convert an electrical signal into physical movement or vice versa. Loudspeakers, for example, are transducers. They produce pressure waves we hear as sound in response to an electrical input. Sensors are also transducers but the other way round: a mechanical input and an electrical output. For example, measuring blood pressure depends on a transducer to sense pressure and transmit the result to a display. Though few realise it, transducers of all kinds lie at the heart of many types of modern electronic equipment. A typical car contains hundreds. Transducers are everywhere. We take them for granted.

Transducer technology is – to put it bluntly – long in the tooth, unglamorous and neglected in terms of development. Unbelievably, the actuators used in many modern electronic goods rely on nineteenth-century technology. There was, and still is, a wide gulf between zippy electronics and clunky mechanics. Founded by an entrepreneurial physicist, Dr. Tony Hooley, in developing its innovative actuator, Cambridge

Mechatronics has created a technology that is compatible with twenty-first-century electronics.

The three consecutive dots in the heading of this chapter are deliberate. Cambridge Mechatronics started life in August 1995 as '1...Limited'. The '1' referenced the unary method of coding digital audio signals that Tony Hooley had invented. At this point, his thought processes were taking him in the direction of a loudspeaker based on unary electronics. A unary number system is the simplest numerical system. The contrast is, of course, with the decimal system and the binary system that computers use. Unary devices are inherently attractive because, being identical, they are easier to manufacture. In the event, it soon became clear that the unary approach was not a runner, and binary was the way to go. But he decided to stick with the name.

Normally, in English, an ellipsis (three consecutive dots) indicates that a word or sentence has been left out intentionally. Not so in this case. Tony told me it was a mathematical in-joke. In maths, an ellipsis after a number or series of numbers means 'and so on indefinitely'. An array of unary transducers would be an array of 1s constantly repeated. By (mathematical) definition, it can't be limited! Whatever the meaning, it served to both intrigue and confuse customers and investors and, sensibly, in 2008, the directors changed the name to the much more descriptive Cambridge Mechatronics. However, outside the world of engineering, this name too has its problems. It is an accurate description of what Cambridge Mechatronics does: the merging of mechanics and electronics. But to a lay person, even an educated one, the meaning of 'mechatronics' is not immediately obvious.

Tony Hooley is the sort who cannot resist tinkering with things to see how they work and if he can improve them. He has a mind that is constantly fizzing with ideas. As a boy, he took valve radios apart. He is a techie of an earlier generation. After gaining a first-class degree at the University of London, he came to Cambridge to do an astro-physics PhD in Radio Astronomy at the world-famous Cavendish Laboratory.

He soon, however, decided that astronomy was not for him and, in 1978, he and a colleague set up Eicon Research Ltd. Eicon sold its services as a consultancy but was soon drawn into becoming an Apple II dealer as that was the machine they were supplying to their customers for their own applications. Noticing that a drawback of these early computers was a lack of storage, Eicon designed and manufactured a high-capacity floppy disk. For a time, the company grew rapidly. Then, in 1986, Eicon ran out of cash and collapsed. In retrospect, Tony admits he made the classic error of becoming too preoccupied with the rapidly developing technology at the expense of the numbers. In his words:

> 'Back then I was the archetypal boffin with some good technical ideas – now I think I know a bit more about some of the all-round issues of running a company … I am very watchful about the company's cash flow. It is very easy in a technology company to keep telling yourself the technology's wonderful but lose concentration on the financial side.' ('A second bite at the marketplace', Peter Marsh, *Financial Times*, 18 October 2001)

Tony decided to go to Australia. Why Australia? He was invited by a schoolfriend who was running a gold mine in Western Australia to spend a holiday with him. He did – and ended up staying for seven years. For much of that time he was doing technical consultancy in electronics, physics and software as well as running his own company in Sydney for three years.

Returning to Cambridge in 1994, Tony had a software idea but had to give up trying to fund it as investors were reluctant to back something that was not capable of being patented. He then came up with the idea of making a fully digital loudspeaker, which *could* be patented. It would produce true surround sound from a single panel. This is effective because of the way the human ear works, correctly perceiving reflected sound as coming from the direction of the last reflection. Until then, it was only possible to achieve surround sound by placing several loudspeakers strategically around a room in a configuration known as 5.1 surround sound, referring to the number of channels the system carries. The '.1' refers to a separate channel that only reproduces low frequencies, requiring the use of a special speaker called a subwoofer, which cuts off frequencies above a set level.

The innovatory nature and commercial potential of this idea were recognised in July 1995 by a DTI (Department of Trade and Industry) SMART Award worth £45,000. (These were government grants to establish the feasibility of inventions and to help develop the product to the pre-production stage.) With this, 1...Limited was born. In setting it up, Tony was able to draw upon what he had learnt in 17 years of running his own business.

In March 1996, David Gillett, an experienced local Cambridge businessman, with whom Tony had a social connection, became his first angel investor. He, in turn, brought in three additional investors. Together they matched the SMART Award, supplementing the drawdown of Tony's savings. David Gillett subsequently became finance and commercial director of 1...Limited. However, by the third quarter of 1996 the money was starting to run out. It was clear that another £100,000 was needed to continue with the digital loudspeaker.

By chance, Tony met a local bank manager, Valerie Dring, who was both familiar with and interested in the high-tech scene in Cambridge. Val set about helping him raise money. She introduced him to a Cambridge-based group of business angels called the Great Eastern Investment Forum. Launched in 1995, GEIF provided access to high-net-worth individuals, local venture capital funds and established companies. Tony was invited to present to them in November 1996. As he was quoted as saying afterwards on the GEIF website:

> 'At the Presentation Day we were delighted to find that, in the first break-out session, our room was full to overflowing with interested investors. In the ensuing month or two, many visited my "labs" at our house, some several times, but by February 1997 we still had not raised a penny. Money was starting to get tight, so we asked the GEIF team to investigate where we had gone wrong. After some research, it turned out that while investors had been intrigued with the digital electronics, they felt we needed to develop a digital transducer to complete the package.' ('Success Stories: 1...Limited', www.geif.co.uk)

GEIF panel members had asked Tony a simple question: 'What transducers are you going to use?' To which the answer was 'I don't know yet'. Truth was, he had been so busy sorting out the electronics he hadn't given any thought to it. Although his key contribution to the idea of an improved loudspeaker lay in the thinking he had brought to bear on signal processing and amplification, he realised that the transducer was potentially as much of a bottleneck. It could be magnetic – the existing technology – but it made sense to consider other, better materials that would produce, via lots of little pressure pulses, sound.

ADDRESSING THE ACTUATOR ISSUE

The objective was a very small, efficient transducer. Specifically, an actuator. As explained earlier, an actuator is a specific type of transducer that responds to electrical signals. The output is always a physical force or movement. Actuators are everywhere in our daily life: for example, the mechanism that unlocks your car doors when you press the button on your remote key. Or the electro-mechanical device that turns your boiler on every morning by opening a gas valve in response to an electrical signal from your central heating timer.

Consequently, during 1997, the focus of attention turned to identifying a reliable and compatible actuator. There are several types of electrical actuator. Magnetic coils are well established in applications like loudspeakers in spite of their disadvantages, which include high power consumption, weight and the fact they necessarily generate a magnetic field. To do what he wanted it to do, the flat panel digital speaker required a large array of identical small transducers. Each of the actuators in the array would be 'driven' separately by a clever piece of signal processing software he was developing. Electromagnetic transducers worked but were unsatisfactory. Six months after the initial presentation, he was back making his case to GEIF members, this time with a firm proposal on transducers.

> '... we presented a second time and were dismayed to find our break-out room half empty. But this time it was quality not quantity, and the investment we needed came rolling in – with our existing investors and other acquaintances chipping in too.' ('Success Stories: 1...Limited', www.geif.co.uk)

With a further £100,000 under its belt from GEIF, 1...Limited (comprising Tony and a student research assistant) was able to focus its attention during the rest of 1997 on the transducer issue. Shortly after the GEIF money came in, 1...Limited received a further grant from the DTI: an £80,000 SPUR award. (SPUR was a parallel scheme to SMART.)

After evaluating several possibilities, Tony concluded that piezoelectric transducer technology offered the best prospects. The expansion or contraction in a piezoelectric actuator is caused by a reversible change in the structure of the material, meaning that the movement can be switched on and off – very rapidly if desired – via an electrical input. In terms of material, he concluded that ceramic offered the best prospects. Certain ceramic materials have intrinsic piezoelectric properties. Equally, ceramic can be manufactured easily into a variety of sizes and shapes for diverse applications. Because there are so many types of transducers for multiple applications (pressure, mechanical movement, temperature, etc.) the industry that makes these is highly fragmented.

As there was nothing suitable available off the shelf, Tony consulted several university experts and, with their input, 1...Limited constructed a ceramic device. Unfortunately, it proved to be deficient in several respects. Consequently, during 1998, 1...Limited embarked on a collaboration with several universities to try to solve the problem. Fortunately for 1...Limited, the European Union had decided to fund research into this area. This collaboration morphed into a European-wide consortium of four universities and four industrial partners, 1...Limited being the technical lead. The objective of the EC-funded research project was to develop and produce a superior ceramic actuator for use in digital loudspeakers and other applications. 1...Limited continued to benefit from the consortium, both technically and financially, for close to four years, receiving £470,000 in EU research grants between 2000 and 2003.

The breakthrough that allowed the group to progress was made in conjunction with the University of Birmingham materials science department. In return for surrendering the patents and know-how, the university was granted shares in 1...Limited. The original ceramic transducer design had several design challenges. By the end of 1999, however, these issues had been overcome with the development of the Helimorph ceramic actuator shown opposite.

1...Limited's design used PZT (Lead Zirconate Titanate), a well-known ceramic material, in an unusual super-helix structure: the Heli in Helimorph, as shown above. The actuators were formed from multiple layers of PZT, which are surrounded and separated by

Source: Cambridge Mechatronics Ltd (CML)

conductive electrodes. Not only did the Helimorph design work, it worked extremely well. The physical movement in response to an input signal was substantially greater than that from any existing piezoelectric transducer technology. In colloquial terms, you got a lot of bang for your buck. And it had other attractive features like low power consumption and silent operation. It was clear that the actuator they had invented had definite potential in a wide range of applications.

By the 1970s, US pioneering efforts in the piezo ceramic field had lost momentum and the focus of development had shifted to Japan. Several Japanese companies majoring in ceramics emerged at that time and attained world leadership. One of these was the broad-based electronic components company, TDK. TDK will appear on numerous occasions in the CML story.

With all the attention on the actuator itself, it is easy to lose sight of the fact that, without its control circuitry, an actuator is like a body without a brain. It needs a set of software instructions embodied in a 'driver' chip to function. A driver integrated circuit (sometimes called a controller IC) is a circuit or component used to control another circuit or component. It exists to manage continuous signal, voltage and current outputs. Just to be clear, the term 'driver IC' in electronics is a generic one. They are used in a wide variety of electronic equipment and devices, everything from computer monitors (to drive the display) to utility meters.

In popular parlance, ICs are microchips, or just chips. In this book, 'ICs' and 'chips' are used interchangeably. Driver ICs are the interface between the transducer and the external world. Suitable components were not available off the shelf so 1...Limited started looking at using a specially developed integrated circuit for this purpose, taking advantage of the rapid advances in IC technology.

BUSINESS MODEL AND FINANCE

Right from the start, Tony Hooley conceived of 1...Limited as an intellectual property (IP) owner that would receive fees from licensing its technology to others. He had no desire to get into manufacture, other than on a laboratory scale to produce working models and samples for potential customers. Having decided to pursue a licensing business model, protecting the numerous patents he had filed was clearly crucial, so Tony recruited Dr Ursula Lenel, whom he knew, to become director of IP. 1...Limited then approached a number of companies with expertise in ceramics to manufacture the Helimorph in quantity.

It is noteworthy that several companies had by this time proved the viability of an IP-based business model. Chief among these was the chip design company, Arm (originally, Acorn RISC Machine), also based in Cambridge, which had been operating on this basis since 1993. Arm's foundational architecture designs for semiconductor processors were not sold for a one-off payment but licensed to customers for an up-front fee and then per-unit royalties linked to each chip shipped. Semiconductors based on Arm's technology had several advantages but chief among these was their power-saving capability, thus offering the longer battery life that portable devices needed to become viable.

By the mid-1990s Nokia, the mobile market leader, had firmed up on Arm's design architecture for its phones. Subsequently, Apple developed a close relationship with Arm and is a key customer for the System-on-a-Chip that powers the iPhone. Arm continues to provide the technological underpinning for almost every smartphone sold. As such, the company was – and is – a key enabler of the mobile revolution. Arm originally went public on the London Stock Exchange in 1998. In 2023 it was reincarnated as a New York listed company. The other well-known IP-based company at that time was Dolby Laboratories, the licensor of the electronic noise reduction system invented by Ray Dolby in the 1960s. By the mid-1990s Dolby technology had been extended to the cinema and was producing a royalty income of $60mn a year.

Between 1995 and 1999, 1...Limited operated from Tony's house. He had converted the downstairs kitchen, toilet and laundry room into a lab, with a sound studio in the bedroom above. In 1998 he recruited a couple of engineers to work with him, all of them working from his home. As the year progressed, it became clear that more funds were needed to secure the future of the digital loudspeaker. He also knew that the time had come for 1...Limited to move into proper premises with scope for expansion.

It so happened that the wife of David Gillett (finance and commercial director) – Penny – was a schoolfriend of Christine Lione, who was then company secretary at Flemings, the private investment bank in the City (which was acquired by JPMorganChase two years later). Christine was aware of 1...Limited and suggested to Penny that Tony should approach Philip Bradley, a director in the corporate finance department, to see if Flemings might be interested in investing. Tony Hooley and David Gillett met Philip at Flemings' offices in the City at the back end of 1998. Philip listened intently for an hour and a half. Nothing was said at the end of the meeting, but Philip felt that what Tony had invented was interesting. It had the advantage that the product was one that any potential investor could readily understand.

Surround sound existed but everyone knew it was cumbersome and left a mass of unsightly trailing wires round the room. He also liked the fact that Tony had acquired commercial nous through his experience of running his own business – including failure – for many years.

The next day Tony got a phone call from Philip, saying that 1...Limited was not one for Flemings' funds but he and some friends were keen to explore how they might invest. He immediately arranged to visit Tony's home office in Cambridge where he was treated to a demonstration of a crude mock-up of the digital loudspeaker system using the components then available. After some ups and downs, the result was a £1mn private placement in September 1999. As 2000 progressed, the network of private investors recruited by David Gillett and Tony Hooley expanded as they brought further individuals into the fold.

In January 2000, the board of 1...Limited concluded that the transducer justified its own dedicated R&D programme. When it became clear that further equity from new private investors would be needed to fund this, Philip Bradley worked his wide range of City contacts and received commitments from a substantial number of individuals who were prepared to invest serious amounts of money. Philip's experience and connections provided much of the energy behind this and the two subsequent rounds of fund-raising.

The result was that, during 2000, a further £3.9mn was raised via a private placement. At that point, the investor base comprised around 50 private shareholders. It is important to remember that these transactions occurred at the height of the stock market dot-com boom, when anything associated with the internet or electronics instantly attracted a sky-high valuation. Private transactions reflected what was going on in public markets. The stock market rose rapidly in 1999 and peaked in March 2000 although values remained high for most of that year. By 2001 prices had fallen substantially.

Why did 1...Limited choose to go down the angel investor route rather than seeking capital from venture capital firms? As Tony tells it, the answer is simple. He had been talking to VCs about investment from the beginning. But the deals that they offered, in terms of the value they were prepared to attach to the company, were never remotely as good as those he could obtain from private individuals. In this, Tony was supported by Philip. As well as the low value venture capital firms were prepared to place on 1...Limited, they invariably insisted on preference shares and pre-emption rights. Bringing them on board would, in his view, add an additional layer of complication operationally. And, in contrast to several of the leading Silicon Valley VCs, it was doubtful whether UK-based venture capital firms could assist with the development and marketing of the technology.

Around the time of the 2000 fund-raising, the small board that existed decided that 1...Limited needed a more formal management structure. Mark Shepherd was recruited as chief technology officer and, in late 2001, was appointed CEO, to allow Tony, who was performing this role, to spend more time on research. Andy Osmant was recruited as finance director in January 2001. Simultaneously, David Gillett brought in some non-executive new blood in the shape of two senior business contacts who had invested in 1...Limited. Philip Bradley also joined 1...Limited's board in 2000, as did Christine Lione as company secretary.

2.

Marketing the Digital Loudspeaker

PROGRESSING THE PRODUCT

The year 2000 was spent perfecting the Digital Loudspeaker (as it was now called). The engineering team had executed a classic pivot, modifying and further developing the digital signal processing and reverting to more conventional but still specially designed magnetic transducers. By October, the 30-strong team at 1...Limited, now located in purpose-built premises at the St Johns Innovation Centre in Cambridge, had constructed a functioning product capable of producing high-quality surround sound in a room from a single slim panel. It did this by controlling eight independent beams of sound and bouncing them off the walls. Existing surround sound systems required a wired connection to each speaker, so this was a real breakthrough.

Traditional loudspeaker manufacture is essentially a matter of woodworking. An electromagnetic motor, in the form of a voice coil round the apex of the speaker, replicates an analog electrical signal as an air pressure wave by vibrating the membrane of the speaker housed in an acoustic box. ('Analog', as opposed to the British 'analogue' is used in this book: it is the spelling commonly used in the electronics industry.) The magnetic coil principle has been known since the 1870s. It has since been extended beyond sound to become the motive force in a wide range of industrial and consumer applications.

Used in this way, a magnetic coil becomes the basis of a transducer called a Voice Coil Motor (invariably abbreviated to VCM), converting electrical energy into mechanical motion. For historical reasons, it continues to be called a *Voice* Coil Motor even though the output is not sound but motion. The continued use of 'voice' in the name is clearly misleading and confusing, but it has unfortunately retained its hold. Equally, it would be more accurate to refer to Voice Coil *Actuators*. To the lay person, the word 'motor' has connotations of an engine of some kind. A better nomenclature altogether would have been Magnetic Coil Actuator, but we are stuck with Voice Coil Motor.

What the Digital Loudspeaker achieved, by employing advanced digital signal processing and techniques borrowed from radio astronomy, was the elimination of multiple separate loudspeaker boxes and their cables. Moreover, it could do things that a conventional loudspeaker could not. Each transducer in the array could be programmed differently, opening up the possibility of being able to direct sound to a particular point such as, for example, a single platform in a mainline train station. Clever software allowed the user to generate and steer up to eight beams of sound simultaneously.

At the beginning of 2001, 1...Limited had a demonstrably functioning product so the next question was: what was the best way to insert the Digital Loudspeaker into the audio marketplace? A Memorandum of Understanding for an advanced version of the current product, envisaging a sound panel adjacent to a flat television screen, was signed with Philips in March 2001. The idea was to use 1...Limited's signal processing technology in conjunction with new, high-performance transducers developed by Philips. For technical and marketing reasons, this product never saw the light of day.

Not wishing to tie itself to a single customer, 1...Limited initiated discussions with other major organisations in the audio industry, including Sony, Yamaha and Harman Kardon. In January 2002, the company launched the MK5 Digital Loudspeaker at the annual Consumer Electronics Show (CES) in Las Vegas. It was well received. The 1...Limited exhibition stand was an invitation-only room. Tony's note written at the time reads:

> '... we have all the big names from the industry here, including Ray Dolby, Sony, Bose, Harman, TI, and hundreds of others. People's jaws literally drop – they are gob-smacked. The Sony Japanese guys are sent to tap the room's walls to find what they believe must be the "hidden" loudspeakers ... their engineers say, "Fan-a-tactic! We all lose jobs!"'

Later that year, 1...Limited invited representatives of all the leading audio companies to a demonstration in Cambridge. It also went well and those who attended appeared impressed.

In the meantime, 1...Limited decided to go ahead with its own-brand version of the Digital Loudspeaker to accelerate awareness of this revolutionary audio concept and encourage interest from manufacturing licensees. The company went public with this in September 2002, calling it the 1Digital Surround Sound Projector to distinguish it from the standard product available for licence (now being called the Digital Sound Projector, or DSP). However, the cost of production – carried out under contract in the UK – was very high, with little prospect of getting it down. At a retail price of £25,000, it had to be pitched right at the top end of the domestic market, such as home theatre, and professional applications. It was hailed by the technical audio press, who expressed the hope that volume production would bring the price down to an affordable level. The 1Digital Surround Sound Projector looked like this:

Source: Cambridge Mechatronics Ltd (CML)

While this product was in the pipeline, 1...Limited signed a licence agreement with Pioneer Corporation of Japan to manufacture and sell the standard Digital Sound Projector. Pioneer asked 1...Limited to build and test the DSPs so as to be able to supply them with a ready-to-sell product. The Pioneer DSP was formally launched in November 2002. There was an understanding by both parties of the urgent need to reduce the production cost if it was to be manufactured in Japan for sale to the global audio market.

To this end, in 2002, 1...Limited began the design of two application-specific integrated circuits (invariably referred to as 'ASICs' in the electronics industry) to act as driver ICs. ASICs play a crucial role in the electronics industry generally, and in the consumer sphere especially. These chips were made to 1...Limited's design with the intention of having production parts ready by the summer of 2003. Using ASICs would 'siliconise' the control software: the sophisticated digital signal processing element plus the ability to handle beam steering. This use of custom-made semiconductor control chips was projected to have a major impact on production costs: a reduction of as much as £2,000 per unit.

PROMOTING THE DSP

Pioneer insisted on several design changes to the prototype low-cost DSP that 1...Limited was developing, which delayed its introduction. Pioneer was under intense competitive pressure at this time, culminating in a huge loss in 2005, leading to job cuts in Japan and elsewhere. The low-cost DSP project became a casualty of these events. Nevertheless, 1...Limited earned over £300,000 in up-front licence payments and royalties from Pioneer in the financial year to 30 September 2002 (and a further £320,000 in the next financial year).

What was 1...Limited's financial position at this point? A presentation to shareholders in the Autumn of 2002 provides a snapshot. Since inception, the company had raised £9.3mn from an investor base that had grown to 160 individuals and a sprinkling of institutional investors. The latest funding round had valued 1...Limited at £17mn. It was, inevitably, heavily loss-making but at least revenues were starting to come in.

Tony and his team talked to a wide range of potential users of the technology incorporated in the DSP, both within and beyond the audio field. These included Dolby, which was interested in developing a Cinema Sound Projector and, in fact, funded two projects with 1...Limited to assess the feasibility of this. In the event, nothing came of these conversations.

One thing that had become increasingly apparent to 1...Limited was that audio companies wanted a ready-made electronics package they could use as the core of a single-panel surround sound offering. By the end of the fourth quarter of 2003, the two ASIC chips had been designed and manufactured and 1...Limited was ready to go to market with just such a package solution. Extensive discussions were held with several well-known Japanese audio companies, including Sony, Toshiba and Yamaha. It was crucial for the future of 1...Limited that one of these initiatives turned into something solid.

3.

Semiconductors: a quick and easy guide

CHIPS AND NODES

It will be evident by now that the semiconductor chip embodying the control circuitry is as important as the actuator. One cannot function without the other. The history of 1...Limited is only comprehensible in the context of the semiconductor revolution. 1...Limited had contracted with a semiconductor firm to design and make the two ASIC driver chips it needed. An ASIC has a specific function in the equipment for which it is designed and cannot be used for anything else. As such, it is able to confer on the equipment unique features that distinguish it from the competition. The customer of the semiconductor company owns the rights to that ASIC and any software embodied in it. This makes it difficult for a rival to 'reverse engineer' the product.

Size really does matter in the semiconductor industry – small size. For the past 50 years, the race has been on to make transistors smaller in order to cram as many as possible onto a single integrated circuit. Progress has been remarkable since Intel's first microprocessor in 1971, which contained a mere 2,300 transistors. Smaller transistors lead to higher chip density, which, in turn, leads to lower energy consumption, reduced cost and higher processing speed.

Semiconductor chips are made in factories that are referred to in the industry as 'wafer fabs' ('fab' for fabrication, not fabulous!). Transistor size is measured in nanometres (abbreviated to nm). A nanometre is one billionth of a metre. The level of density achieved is referred to as a process technology node: the size of the smallest feature a wafer fab is capable of creating on a silicon wafer. Increasingly, in daily usage, the word 'node' is dropped and industry participants simply refer to, for example, a 10nm process. Or, alternatively, it is referred to as a 10nm node. Node is a rough and ready measure of a wafer fab's level of technical capability.

For the majority of industrial applications, a process technology node of around 30nm or above is sufficient. These are frequently referred to as mature nodes. 'Mature' is not, as sometimes, a euphemism for aged and incapable. Quite the reverse. Older, higher node chips are the workhorses of the economy, as is clear from the table below. Media attention is – understandably – firmly fixed on innovation at the glamorous end of the node spectrum. The most advanced chips, like the Apple A19, have transistors with features as small as an

PROCESS NODE LEVEL RANGE	TYPICAL APPLICATIONS
Large node size range: 180–90nm	Automotive electronics, industrial control systems
Intermediate node size range: 90-32nm	Gaming consoles, digital cameras, networking equipment, wireless communication devices, automotive infotainment systems
Smaller node size range: 32-7nm	High-performance computing, data centres, cloud computing infrastructure, AI accelerator chips, advanced networking devices
Leading-edge node size range: 7nm and below	Top end ('flagship') smartphones, high-end GPUs (Graphics Processing Units), advanced AI, autonomous vehicles, augmented/virtual reality devices, scientific research

Source: Tony Golding, adapted from 'Semiconductor 101', Abhishek Singh, Medium, 2 July 2023 astonishing 3nm. The aim is to get the node down to 2nm and lower. These ultra-high performance sub-10nm chips offer advantages, in smartphones, computers and AI data centres especially. There appear to be no hard-and-fast rules on the classification of process nodes by end use. Generally, the table opposite, reflects the consensus.

According to the American trade body, the Semiconductor Industry Association, around 60 per cent of semiconductor chips are used in computer or communications products. Functionally, chips can be classified as follows:

Digital chips

» Digital integrated circuits are either logic or memory devices. Logic chips are designed to perform complex logical operations and execute instructions. They are used in a wide range of applications, in computers and smartphones especially.
» Memory chips are primarily responsible for storing and retrieving data. Widely used, they play a crucial role in providing the system with the ability to retain information temporarily or permanently.

Analog chips

» Analog chips are designed to process and transmit analog signals – continuous electrical signals that can vary over a range of values. Analog ICs connect logic chips to real-world functions such as speed, sound and electrical current. They are used in a variety of applications, including audio and video equipment, sensors, and other types of electronic devices.

Mixed-signal chips

» Mixed-signal ICs contain both digital and analog components and are designed to process both digital and analog signals. Like analog chips, mixed-signal ICs act as the interface between digital electronics and the outside world. Typical applications include mobile phones and medical devices. The driver chip that an actuator needs to function is a mixed-signal IC.

THE MANUFACTURING PROCESS

Chips are constructed on a wafer (also called a slice or a substrate). Hence 'wafer fab'. A wafer consists of a thin slice of semiconductor material, normally silicon. It serves as the substrate for the electronic devices (transistors and other components) that are built on it. Wafer fabs are defined by the diameter of wafers that they are tooled to use. Diameters have gradually increased to improve throughput and reduce cost, with a state-of-the-art fab using 300mm wide wafers.

The manufacturing process begins with design. Once it has been determined what the chip is intended to achieve, the designer specifies the layout with all the interconnections between the components. Design tools take the description of a proposed circuit function and assemble the circuit elements that implement that function. This is a demanding task with many iterations. Today the process is highly software-intensive, making use of a set of tools that are collectively known as Electronic Design Automation (or EDA for short). EDA tools are also used to verify that a design will meet all the requirements of the manufacturing process.

Fabricating semiconductor chips is an extraordinarily complex process, operating at the borderline of physics and chemistry. It comprises a sequence of photolithographic and chemical processing steps during which electronic circuits are gradually created on a silicon wafer. Photolithography – or often just lithography – involves using light to transfer a pattern onto a layer of light-sensitive material called a photoresist on the surface of a wafer. Fabrication takes place inside a cleanroom located in the centre of a wafer fab. Production is completely automated and carried out in a hermetically sealed nitrogen environment to improve yield (the percentage of microchips that function correctly on a wafer at the end of the process). Yield is a critical parameter in semiconductor manufacture. Automated material handling systems take care of the transport of wafers from machine to machine. Manufacturing a finished semiconductor wafer takes 12 weeks on average but can take up to 20 weeks for advanced processes. To ramp up production yields to satisfactory levels takes more time. The diagram below illustrates the sheer number of iterations that a silicon wafer goes through after it has entered the wafer fab.

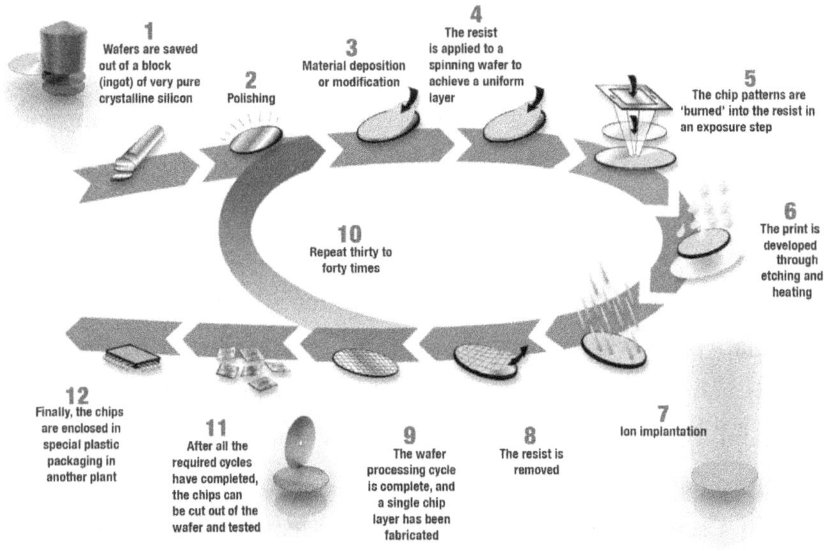

Source: Bits&Chips (bits-chips.com)

Wafer fabs require a constant flow of inputs to function. These inputs are many and varied: raw materials, chemicals, gases and specialised equipment. The sources of these inputs tend to be geographically concentrated. For example, 75 per cent of the world's supply of high-end photoresists – resins used to etch circuit patterns on wafers – comes from four Japanese companies. Some suppliers need to be geographically close to the fabrication facility. The key point is that wafer fabs are reliant on an extensive support hinterland.

At the end of production, the semiconductors on the silicon wafer go through assembly, test and packaging (frequently referred to as ATP) before delivery to the customer. Packaging, as this entire 'back-end' process is often called, is traditionally a deeply unglamorous, contracted-out, low-tech activity. The companies that offer this service are often grouped together in industry parlance as OSAT providers, OSAT being an acronym for Outsourced Semiconductor Assembly and Test. Most OSAT activity takes place in Asia, with Taiwanese and Chinese companies in the lead.

Until recently, semiconductor processing has been the beneficiary of Moore's Law. Gordon Moore, one of the founders of Intel, predicted in 1975 that the number of components in an integrated circuit would double every two years. Astonishingly, his prediction has held good for 50 years. Now, however, Moore's Law is coming up against its limits. As semiconductor fabricators find it harder to squeeze more transistors onto a chip, the focus has shifted to high-density packaging as a means of enhancing performance. This is known as advanced packaging. It involves sophisticated tools and cleanrooms akin to those used in a wafer fab. Consequently, participation is not cheap. The cost of an advanced packaging facility can run to several billion dollars.

In a sense, advanced packaging has become the new technological frontier. Packaging multiple chips more closely together enables chipmakers to increase speed and efficiency. In one popular approach, high-performance ICs known as 'chiplets' are stacked three-dimensionally in a single package, utilising advanced interconnection techniques. Chiplets can be designed, manufactured and tested independently of each other, and then, like LEGO, interconnected to form a larger and more complex integrated circuit. Advanced packaging is employed in mobile phones but is being driven by the requirements of AI, which calls for rapid data exchange between computing and memory elements. With the diminishing influence of Moore's Law, the leading semiconductor fabricators have invested in advanced packaging in an effort to retain control over as much of the value as possible, thus invading the space previously occupied by OSAT providers. In practice, these two sets of back-end operators co-exist.

Excess heat is an increasingly serious problem in electronics. When in use, semiconductors generate a high level of thermal energy. Such is the load placed on smartphones by faster processors, heavier apps and complex graphics that they can generate exorbitant heat. Overheating degrades performance and can crash the phone entirely. To counter this problem, various techniques are employed to dissipate the heat emanating from integrated circuits. Modern semiconductors in high-end phones commonly use a liquid cooling technology known as vapour chamber cooling.

The following diagram endeavours to put the key elements in semiconductor manufacture into perspective. The fabrication box is larger – though less than proportionately – because it embodies the highest level of cost, skill and technological sophistication.

The total semiconductor process

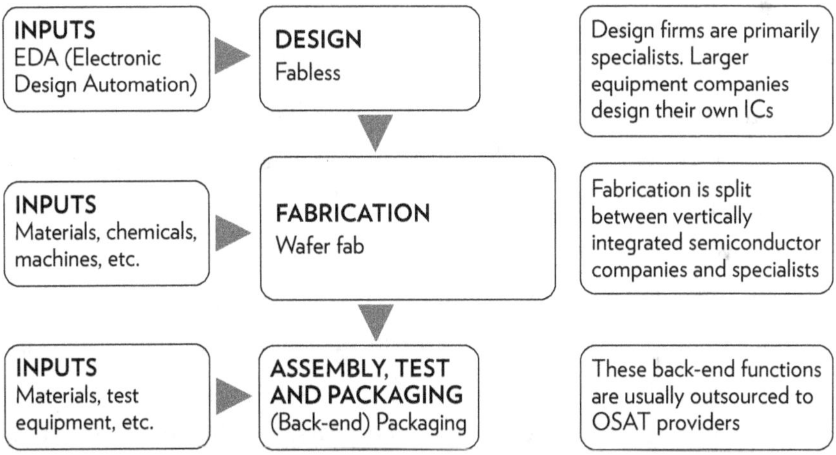

Source: Tony Golding

INDUSTRY STRUCTURE

When semiconductors first emerged in the 1970s the industry had a very simple structure. It is now complicated. In the early days, every firm performed all three functions of design, fabrication and packaging within the corporation. Firms that incorporate all three functions (the central spine in the diagram) are known as integrated device manufacturers (IDMs). Due to the scale of the investment now required for fabrication, few IDMs remain. Prominent examples are Samsung, Intel and Texas Instruments.

Today the majority of semiconductor companies confine themselves to design and then outsource fabrication and packaging. Firms whose *raison d'être* is design are commonly called in the industry 'fabless' semiconductor companies. Put simply, a fabless semiconductor company is one that designs integrated circuits. 'IC design house' would be a much better term. But fabless is what everyone uses. Fabless is an odd description as it implies these companies have lost something – which is true in the sense that originally all semiconductor firms both designed and made ICs. The salient point is: nearly all semiconductor firms design but very few manufacture their own designs. The demand for design services continues to grow as more and more ASICs are specified by equipment makers to differentiate their products from the competition. Larger electronics companies, such as Apple, wanting to create ASICs have set up an in-house IC design function.

The proliferation of chip requirements and the increasing demands of manufacture have led to a division of labour. Design and fabrication skills have little in common, but the real problem is the burgeoning cost of a wafer fab. Consequently, a gulf has emerged between these two activities. Semiconductor fabrication is a gargantuan undertaking, involving tens of billions of dollars, a very high level of technical competence and a timescale of several years. A fabrication firm that focuses solely on producing ICs to the designs of others is known as a silicon foundry, echoing usage in the metal industry as a facility that casts objects to order. The reason for the existence of this service is the absolutely staggering price of a state-of-the-art, sub-10nm process node, fabrication unit: typically, $12–30 billion. Building and operating such a facility also requires a critical mass of high-level engineering expertise. It therefore makes sense for some semiconductor firms to specialise in just making chips, leaving the design function to others.

As the cost of building and operating production facilities has mounted and the level of expertise required has advanced, the foundry business has become increasingly concentrated and is now dominated by one company, Taiwan Semiconductor Manufacturing Company (TSMC), which controls more than 60 per cent of the global contract (foundry) market. Samsung ranks a distant second. Nor is it just a matter of money. TSMC has an accumulated engineering skill base in Taiwan that others find difficult to replicate. Equally, it is a financial powerhouse with, in early 2026, a market capitalisation of $2 trillion, making it the sixth most valuable company in the world.

Foundry contenders are therefore few in number. More to the point, only the two leaders – TSMC and Samsung – are capable of operating at leading-edge process nodes below 10nm. However, since 2022, Samsung has been unable to match the yield levels achieved by TSMC. Similarly, Intel, the premier American semiconductor company, has struggled since 2020 to match TSMC's manufacturing acumen. Consequently, neither has been able to dent TSMC's overwhelming lead.

Many of the leading fabless semiconductor companies are dependent on TSMC for fabrication of their self-designed ICs. Noteworthy examples are Apple and Nvidia. TSMC and Apple enjoy a symbiotic relationship. The key ASIC in the iPhone 17 relies on a 3nm TSMC process. By the same token, as it accounts for close to one-fifth of

TSMC's sales, Apple effectively underwrites the company's cutting-edge developments.

There is one important segment of the semiconductor industry that is an exception to the general rule on structure and remains vertically integrated, with the design, fabrication and packaging functions all in-house. The dynamics of memory chips – accounting for around a quarter of global semiconductor sales – are completely different from the rest of the industry. Memory chips are commodity items of standard design. They do not require cutting-edge technology – typical process nodes are 10-20nm. There is even a spot market for the most common memory ICs. As such, they lie at the opposite end of the spectrum from ASICs. Competition is all about price and only those producers capable of fabricating at huge volume and low cost in dedicated wafer fabs have been able to stay in the game. The business is consequently the province of a handful of companies, all specialists in this field, the single exception being Samsung.

TSMC has high visibility, leading to the assumption that it is the largest semiconductor manufacturer worldwide. In fact, it is not. Estimates suggest that foundries in total account for around a quarter of aggregate semiconductor output. Using the metric of wafer fab capacity, Samsung is the global corporate leader as it is involved in every sector of the industry, including a major presence in the high-volume memory business. According to Knometa Research Corp., a consulting firm that maintains a database of semiconductor facilities, in 2024 Samsung operated wafer fabs accounting for 19 per cent of global capacity. TSMC came second with 13 per cent.

In the media and generally, the term 'semiconductor industry' is employed with great imprecision. Journalists and others tend to assume that the outsourced (foundry) portion of the industry represents the total. It does not. The industry is characterised either as all the firms that design semiconductors – ignoring the inconvenient fact that most do not possess the facilities to fabricate their own designs – or it covers only those firms that operate wafer fabs for hire (foundries). Both are wrong. In summary, the industry structure – other than the memory sector – is best characterised as follows:

- **Designers of semiconductors.** These fabless semiconductor firms possess an engineering design capability. The resulting ASIC designs are then turned into chips by contract fabricators (foundries). There are many small firms offering a design service to electronic equipment producers. Some producers, like Apple, engage in their own design. A leading semiconductor design house is Qualcomm, a little-known American corporation. Like Apple, it uses TSMC. Unlike Apple, Qualcomm sells the resulting ICs commercially. Qualcomm will figure frequently in this narrative as it occupies a pivotal position in the smartphone industry.
- **Integrated device manufacturers.** IDMs own wafer fabs, and design and fabricate standard ICs for sale on the open market. Texas Instruments, the leader in analog ICs, is a good example, as are the handful of memory chip producers. A large vertically integrated firm like Samsung is involved in all aspects of semiconductor activity, including a foundry service. In 2021, Intel joined Samsung in pursuing the foundry route.
- **Contract fabricators (foundries)** providing an outsourced manufacturing service to designers of ASICs (as above). This sector of the semiconductor industry contains relatively few companies due to the enormous cost and exceptionally high level of skill involved in running a contract wafer fab. Foundries will make to order at any node, but TSMC and Samsung represent the leading edge of fabrication in their ability to manufacture sub-10nm chips. The third ranking company globally is Semiconductor Manufacturing International Corporation (SMIC), the largest Chinese foundry service provider. Just behind SMIC is UMC,

which, like TSMC, is based in Taiwan. TSMC alone absorbs 12 per cent of Taiwan's entire electricity output and semiconductor production in total contributes 15 per cent of the island's GNP. In fifth place lies the US-headquartered, Abu Dhabi-owned firm, GlobalFoundries.

CHIPS AND GEOPOLITICS

Taiwan's pre-eminent position places computer and smartphone customers in an invidious position. To maintain their leadership, companies like Apple and Nvidia must have access to the most advanced chip technology. This inevitably has major economic, strategic and geopolitical implications, with firms having to place their trust in a cluster of wafer fabs in north-west Taiwan – not a comfortable position to be in. As the country where the leading chip design firms are located, the United States is acutely conscious of its vulnerability to the oft-stated Chinese ambition to 'reintegrate' Taiwan.

The prevailing landscape was neatly summarised in a 2021 Boston Consulting report. The situation has not changed materially since then.

> 'About 75 per cent of semiconductor manufacturing capacity, as well as many suppliers of key materials – such as silicon wafers, photoresist, and other specialty chemicals – are concentrated in China and East Asia ... all of the world's most advanced semiconductor manufacturing capacity – in nodes below 10 nanometres – is currently located in South Korea (8 per cent) and Taiwan (92 per cent).' ('Strengthening the global semiconductor supply chain in an uncertain era', Boston Consulting, April 2021)

This quote, while accurate in itself, overstates the actual balance of power in the semiconductor industry. The USA may lag in chip fabrication, but it occupies, together with Japan and the Netherlands, a powerful position upstream in EDA tools and in the manufacture of critical chemicals and production equipment. Chip manufacture in Taiwan or China or anywhere else is dependent on access to these inputs.

Ultimate control therefore resides with a clutch of American, Japanese and European companies, headed by the Dutch company ASML, which is the sole global manufacturer of extreme ultraviolet (EUV) lithography machines, essential for the manufacture of sub-5nm chips. Lithography employs light to transfer intricate patterns onto a photosensitive material (photoresist) on a silicon wafer. EUV machines are so complex that they are well-nigh impossible to replicate. Mature node chips use less sophisticated deep ultraviolet (DUV) equipment. Together with the Japanese companies Nikon and Canon, ASML dominates the market for DUV machines.

One of ASML's key strengths lies in its collaborative approach of working with leading specialist sub-contractors rather than building everything in-house. The EUV machine pictured below, assembled from 100,000 parts sourced from all round the world, is the size of a double-decker bus and costs $200mn.

Source: ASML

ASML is easily the most valuable European company. During 2024 ASML began deliveries (to Intel, TSMC and Samsung) of a supercharged version of its EUV machine, at a cost of $350mn apiece. As the sole global EUV manufacturer, ASML represents a chokepoint in the wafer fab supply chain. Consequently, with the co-operation of the Dutch government, the United States has leverage over wafer fabrication, which it is fully prepared to use.

It is both interesting and significant how the US semiconductor industry sees itself. Every year the Semiconductor Industry Association publishes a Factbook. The 2025 edition proudly announces near the head of the report that 'The US semiconductor industry has more than half of global market share'. This is disingenuous. True, the USA has undisputed leadership in design plus software and production tools, but it plays a minor role in fabrication. At no point does the Factbook mention the unpalatable fact that the vast majority of US semiconductors are fabricated offshore. Only in May 2024 did the SIA admit, buried deep in a press release, that the USA hosted only 10 per cent of world chip manufacturing capacity.

Reassuringly, the Biden administration did not share the SIA's rose-tinted view. The CHIPS and Science Act, passed in 2022, allocated substantial funding to all aspects of semiconductors, including major incentives for the expansion of fabrication facilities on American soil. Intel, as the de facto national semiconductor champion, was the largest recipient of CHIPS Act funding. TSMC responded to the concerns expressed at this degree of concentration in fabrication capacity with a commitment to construct three wafer fabs in Arizona for a combined cost of $65bn. The first of these started small-scale production of chips for Apple late in 2024. Samsung too is expanding its existing American semiconductor production capacity, with an additional fab in Texas due to start production in 2026. All of these new plants have experienced delays. Progress has been constrained by the worldwide shortage of the specialised engineers and technicians required to both build and operate them and, simultaneously, the need to create the necessary support hinterland.

President Trump had consistently criticised the CHIPS Act and, on arrival at the White House in January 2025, replaced the section of the Commerce Department that administered it with a new entity focused on encouraging – rather than subsidising – the construction of semiconductor facilities. Following the imposition of tariffs, TSMC added a further three wafer fabs (plus packaging and an R&D centre) to its US plan, at an additional cost of $100bn. The company also opened a facility in Japan in early 2024 and a second one is under construction. In Europe, it broke ground for a wafer fab in Germany during 2024 in a joint venture with three local companies. It should not be assumed from TSMC's expanded commitment that the company has been persuaded to shift its centre of gravity to the USA. Wafer fab capacity will continue to increase in Taiwan. The company envisages that possibly 30 per cent of its cutting-edge chips will be made in America by 2030. Taiwan will continue to be the location of TSMC's most advanced production technology and its core R&D.

ASICs

Application-specific integrated circuits are big business in the semiconductor industry. They account for, by value, about half the chips that go into smartphones and two-thirds of those used in consumer electronic products. Specifying, designing and producing ASICs is a complex multi-stage process with a substantial up-front cost. Normally, the equipment maker contracts with an ASIC semiconductor design specialist and works with the firm to create a manufacturable design. As we have seen, some semiconductor firms are able to do all of this in-house. But the great majority of ASICs are designed by firms that only design and then send the final version to foundries on behalf of the equipment customer. It is akin to specifying the toppings and exactly how they are laid out on a pizza and then taking it to someone who has a fully featured, professional pizza oven and knows how to use it.

The ASIC design and manufacturing cycle is a long and complex one. The design phase typically takes 6–18 months. The conclusion is 'tape-out' (so called because it harks back to a practice in the very early days of electronics), which is the point at which the photomask of the circuit has been fully created and is sent to the semiconductor manufacturer. Preparing the design for production occupies a further three to six months. All in all, designing an ASIC ready for production takes between one and two years: the cost is typically $1–3 million. For a startup like 1...Limited, the two ASICs it needed represented a substantial up-front cost as well as involving a considerable lead time.

As semiconductor technology advanced, it became possible to pack more and more components onto a single IC: the System-on-a-Chip to which we referred to earlier. The SoC is the pre-eminent chip – among a tally of two dozen or more – in a smartphone and occupies a comparable role in many other types of electronic equipment. Creating an ASIC to perform this function is a significant undertaking, which is reflected in the cost of the chip. An SoC for a higher end, flagship phone can be priced at up to $200 per unit and can represent as much as 25 per cent of the total 'build cost'. What exactly an SoC does and how it relates to the other components in a handset is covered in Chapter 7.

4.

The Yamaha years

SOARAWAY SOUNDBARS

The outcome of 1...Limited's negotiations with the Japanese audio companies during 2003 was that Yamaha was the one to come up with an acceptable proposal. Their engineers had been working for years to find a way to make a thin, single-panel surround sound speaker but had hit a brick wall. Yamaha Corporation's core business was then – and is now – musical instruments, in which it is a world leader. It entered the audio market in 1954, at the same time as it spun off its motorcycle division, as music and motorcycles do not make natural bedfellows.

Within Yamaha there was real commitment to create a low-cost Digital Sound Projector (DSP) and events moved fast. The company pulled out all the stops to show a prototype at the Consumer Electronics Show (CES) in Las Vegas in January 2004. It attracted a lot of attention. Following this, Yamaha set a goal of mass producing a unit for launch by the end of the year. It was a tight timetable but the YSP-1 Digital Sound Projector, as they had branded it, emerged in December. By May 2005 it was available in the USA and the rest of the world. The YSP-1 consisted of an elongated metal box (as shown below) designed to sit below a large-screen television.

In parallel, 1...Limited continued to pursue other licensing opportunities. Management talked to Sony, who were trying to

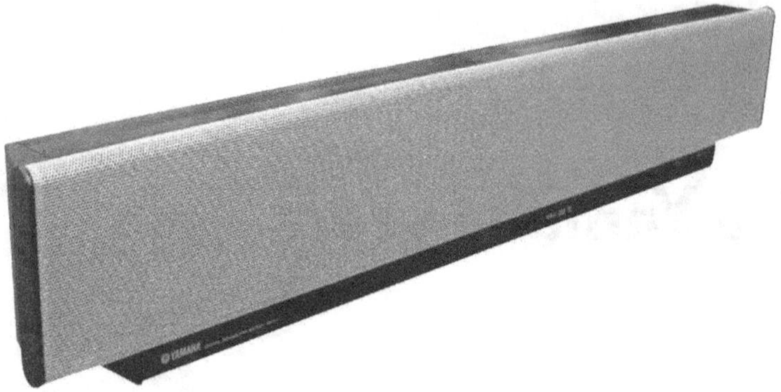

Source: Tony Golding

develop their own surround sound product. Meetings were held with no resolution. Sony was one of several companies that had tried to turn surround sound into a mass consumer product and failed.

Inevitably, taking its cue from the long, low shape, the audio industry started to refer to the YSP-1 and Yamaha's subsequent models generically as a soundbar. From then on, *anything* that looked like the YSP-1, has been called a soundbar, ignoring the significant differences in functionality and performance. Yamaha retains the trademark Digital Sound Projector. But in industry parlance, and in public perception, all subsequent models are just 'another soundbar', even when they are far more than that. This quote from a specialist hi-fi magazine, a decade after the launch of the YSP-1, underlines the extent of Yamaha's – and 1...Limited's – achievement.

> 'In 2005 no one knew they needed a soundbar or really believed there was an alternative to bulky 5.1 surround systems. It took five years or more for the genius of the YSP-1 to resonate with the world at large. So far ahead of the curve it was barely a gentle incline, the YSP-1 set standards that rival manufacturers still struggle to replicate'. ('100 Best Gadgets Ever: The Connected Age', *Stuff*, 22 July 2014, Simon Lucas, What Hi-Fi?)

At the Yamaha results presentation to investors in May 2005 the company referred to 'robust orders for the YSP-1 digital sound projector'. In 2006 it reported that 'high expectations for the YSP-1 were met'. By this time, the company had introduced a second, cheaper model, the YSP-800. It was designed to complement the 32-inch televisions that were becoming popular. By the beginning of 2008 Yamaha had created a family of seven YSP models and reported that shipments were running at more than 100,000 units annually.

Some of the boom that Yamaha was experiencing could be attributed to a fortuitous circumstance in the last quarter of 2007. Flat panel TVs had been around for several years and had made inroads into the sales of traditional cathode ray tube televisions, but the price was prohibitive for most consumers. Suddenly, in the run-up to Christmas, the price of LCD (Liquid Crystal Display) flat screen televisions started to fall rapidly. When the figures for 2007 were tallied, it was clear that LCD had become the dominant technology and sales subsequently took off. The years 2008 to 2010 were particularly fruitful years for flat screen TVs.

Flat panel TVs have minimal room for loudspeakers. These are usually located on the back or underneath – neither is an ideal position. Soundbars offered the ability to restore speaker quality to TVs, especially bass response. Most TV buyers were simply concerned to improve sound quality rather than actively seeking to install surround sound. Of course, a dedicated minority of hi-fi enthusiasts appreciated the difference between what the YSP range offered and the performance of standard soundbar offerings.

Competition intensified in 2008. Samsung launched a soundbar at CES in January of that year. This quote from an American specialist audio/visual magazine nicely captures the scene at the time:

> '2009 was the year for soundbars. People who don't want crappy TV speakers look to soundbars as thin, affordable solutions. The

> Yamaha system is the best we've heard to date and it is loaded with features. Even for the jaded audio/video enthusiast, there is likely a room in the house for a Yamaha soundbar somewhere.' (*Home Theater Review*, 7 December 2009)

'Thin, affordable solutions' may well have been inferior – as this quote implies – but were clearly 'good enough' for less discerning consumers. As the number of alternatives increased and prices fell, Yamaha was hampered by its positioning at the top end of the market, as indicated by the following quote:

> 'For decades now, the entire range of Yamaha audio and video gear, including hi-fi stereo music and home theater systems, have been thought of as the very best gear coming from Japan, and among the very best gear made anywhere in the world.' (Speakergy, Editor's Choice, 18 February 2021)

By 2012, further major names like Philips, Sony, LG and Harman Kardon had joined the fray. Yamaha continued to brand its offering as a Digital Sound Projector but, to the average consumer, it was simply 'another soundbar'.

INTEGRATING THE DSP

During 2007 several audio equipment producers had approached 1...Limited with the idea of developing an integrated television with a built-in soundbar. By 2010 the DSP could be found in flat panel TVs made by Mitsubishi, Grundig, Loewe and Beko. A critical element in achieving penetration of the major consumer electronics companies was the ease with which they could incorporate DSP technology into new products. To this end, 1...Limited teamed up with Micronas, a

German semiconductor company specialising in designing ASICs for consumer electronics to implement 'a turnkey solution via software and hardware' for audio equipment producers. The 2007 Micronas annual report commented specifically on the DSP:

> 'The Digital Sound Projector is a pathbreaking technology which allows lifelike playback of cinema-quality surround sound. The Micronas platform for this technology is the MSP-M audio chip, which enables TV manufacturers to offer their customers the pleasure of perfect home cinema sound reproduction.'

It was singularly unfortunate for 1...Limited that things started to go wrong at Micronas just at this moment. In a TV market dominated by flat panel screens, Micronas could not maintain the market position it held in cathode ray technology, and its consumer division became heavily loss-making. Micronas decided to close or sell the division. Faced with these travails, in 2008, CML (as 1...Limited had become) was forced to change tack and begin conversations with other ASIC suppliers. It was a setback. But, by July, CML had signed up three semiconductor companies and subsequently began to win television business.

The original licence agreement with Yamaha granted the company exclusivity for stand-alone DSP products. As this agreement was due to end in June 2010, CML began a series of conversations with other parties in the audio field who might offer more to them as a partner than Yamaha. Then, in 2011, intense competition in the integrated flat panel TV market forced the last two CML licensees, Mitsubishi and Beko, to withdraw. The decision was made to extend the licence agreement with Yamaha. Although top of the line in quality, the DSP was expensive to produce. The key problem CML's engineers encountered in trying to cut the cost was that, with a 64-transducer array, the quality was excellent but, as you took out transducers to make it cheaper, the sound quality declined. By the end of 2011, only Yamaha was actively working with CML on new products.

The DSP had undoubtedly been a success for Yamaha but the impression from the reported results in 2009 and thereafter is that the 'bloom was off the rose'. Despite this, CML actually had a bumper year in 2009, with a record turnover from the DSP of £3.2mn, although this included a large one-off licence payment. CML actually declared a pre-tax profit in that year of close to £1mn, an outcome that can also be attributed to urgent measures taken the previous year to cut the cost base. Meanwhile, much earlier, in 2001, it had become apparent to the management of 1...Limited that the Helimorph actuator had scope for application beyond the audio field. Funded by the cash from the audio business, increasing engineering resource had been devoted to developing the technology further and assessing its wider potential.

In 2012 the company decided to sell the audio business. Yamaha was seen as the most likely buyer, but others were approached to check the value of the intellectual property in the marketplace. The deal was actually done with Yamaha in 2013, for a price of £1.4mn in cash. Over the period of the agreement, from 2002 to 2013, CML received a total of £13.75mn in licence payments and royalties from the DSP technology it had developed and nurtured. Yamaha has continued to sell Digital Sound Projectors.

How should we assess the DSP project overall? There can be no doubt about the technological feat it represented. Others, like Yamaha and Sony, with far greater resources, had tried and failed to do what 1...Limited had done. Commercially, it was equally a success, especially for Yamaha, which had shipped an estimated £1.5bn worth of 1...Limited-based products. The company was now free to concentrate on turning the actuator technology it had invented en route to the DSP into a viable business beyond audio.

PART 2
TARGET: SMARTPHONES

Focus on autofocus

CAMERA PHONES

The Helimorph, with its advantages over existing actuator technologies, offered features that were particularly attractive to makers of portable consumer devices. A prospectus issued by 1...Limited to shareholders in April 2001 ahead of a proposed fund-raising (with 'Disruptive Technology' blazoned across the front) listed a dozen potential mass-market consumer applications. Management began to envisage the possibility of inserting the Helimorph into a range of electronic consumer products. Strategically, the view began to gain ground that the function of the DSP was to provide cash to fund Helimorph development. In a private placement document issued in 2001, the Helimorph was given top billing in the section on opportunities for the company – ahead of the DSP.

To provide actuator sample kits to potential customers and to explore techniques for mass production, 1...Limited set up a ceramics processing pilot plant in Cambridge. In parallel, the company talked to several ceramics suppliers. At that stage, the most promising opportunities, as the engineering team saw it, lay in four application areas: optical systems, miniature speakers (for mobile phones and laptop computers), relays, and loudspeakers in consumer products.

By mid-2002, these four possible areas of focus had been whittled down to two: optical systems – for which read 'cameras' – and loudspeakers. It soon became clear that the really big opportunity lay in the nascent market for camera phones. Mobile phones were being churned out in huge numbers and the prospect of a Helimorph actuator being used to move the lenses of the cameras that were beginning to go into them had a powerful appeal.

To this end, 1...Limited set up a joint development with STMicroelectronics, a large Franco-Italian, US-quoted, semiconductor company headquartered in Geneva with a stake in the supply of components for cameras. The specific objective was to explore the possible usage of the Helimorph actuator for autofocus and zoom inside the tiny cameras that were just starting to appear in mobile phones. Nothing came of this particular project, but it turned out to be the first of many contacts with STMicroelectronics.

At this point, the vast majority of mobile phones had screens but no cameras. Sensibly, anticipating cameras would be built into phones, Tony had already patented the use of the Helimorph in an optics role, for the purpose of focusing the lens automatically. Phones with cameras had emerged a few years earlier, the first being released in Japan in May 1999. The first genuine mass-market camera phone was a Sharp J-Phone model, launched in Japan in November 2000.

What did the camera landscape look like before camera phones arrived on the scene? And, specifically, what was the actuator mechanism that drove the autofocus system in digital cameras? Digital cameras, incorporating an image sensor coupled with computer-type storage media, overtook traditional film cameras in the 1990s. A camera image sensor is a device that takes the light entering the camera through the lens and produces a digital image from it. The surface of a sensor contains millions of pixels that capture the light. 'Pixel' takes its name from a combination of 'picture' and 'element'. It is the smallest unit of a digital image or graphic that can be displayed. The total number of these on a sensor is measured in megapixels (MP). Generally, the higher the number of megapixels the higher the resolution (and the sharper the display). But, as is explained in Chapter 20, this rule is far from being universally true.

It is important to appreciate the extent to which the digital camera was a break with the past, a break that led to the speedy demise of Kodak. The camera had evolved into, in essence, an electronic device. Essentially, an image sensor is the functional equivalent of coated celluloid film: both capture the image that enters via the lens but in completely different ways. Image sensors are made using semiconductor techniques in wafer fabs.

In the early 2000s digital cameras were in high demand. Global sales were rising at an exponential rate, from 40mn units a year in 2000 to a peak of 120mn units in 2010. At that point, camera smartphones, with improved cameras inside them, started to be sold in significant numbers and digital camera sales consequently fell off a cliff. The digital camera industry was – and still is – completely dominated by Japanese producers. Today, a much-diminished camera industry is overwhelmingly the province of two companies: Canon and Sony. Nikon, Fujifilm, Panasonic and a few others bring up the rear.

AUTOFOCUS OPTIONS

A digital camera autofocus (AF in the jargon) optical system uses a sensor, a control system and a motor to focus on an automatically or manually selected point. The data from the sensor is sent to an electro-mechanical system – an actuator – that adjusts the lens. In simple terms, autofocus involves the following steps:

1. Light comes into the camera.
2. Light hits the autofocus sensor.

3. The AF sensor in conjunction with the drive circuitry determines if the scene appears in focus.
4. If it's not in focus, the sensor tells the motor to move the lens in the direction it 'thinks' will bring the scene into focus.
5. The lens is moved by the AF motor.
6. The AF sensor tells the lens to stop focusing when the object becomes focused.

These rapid adjustments occur in a fraction of a second until the point is reached where the picture is in focus. Sensors use either active or passive techniques to obtain the information they need. Both active and passive autofocus systems require a computation ability, in the form of an ASIC, to instruct and drive the actuator that moves the lens.

Most digital cameras rely on a motor that is built into the lens unit, an approach pioneered by Canon in the 1980s. One of two actuator technologies is used: ultrasonic or, more usually, an electromagnetic stepper motor. Early phone designers naturally turned to the camera technology they knew in digital cameras as a source of inspiration. However, it quickly became apparent that the motors used in digital cameras were not transferrable, being physically large, relatively expensive, mechanically complex, noisy, slow and power-hungry. So the hunt was on for an alternative.

What subsequently became known as the camera module industry began to emerge in response to handset manufacturers' requirement for a plug-in, integrated camera solution. The camera module subsystem incorporates a lens unit, an image sensor, a lens motor, an ASIC drive chip and associated components, such as infrared filters (needed because image sensors can 'see' infrared light, which would distort the view of the subject). Mobile phone makers were looking to procure a complete sub-assembly they could simply slot into the phone as it moved down the assembly line.

The two technological candidates vying for the autofocus motor function in camera phones were piezoelectric and Voice Coil Motor. At this point in the early 2000s, piezoelectric was viewed as an interesting but still emerging technology. By contrast, VCM was familiar and well established, not least because of its widespread use in audio and many other applications. Large-scale production capacity already existed, which meant that it was cheaper than the piezoelectric solution.

VOICE COIL MOTOR AND ITS DRIVER IC

The basic components of a Voice Coil Motor are a coil of wire carrying a controllable electric current and a permanent magnet. A permanent magnet is one that retains its magnetism. The actuator consists of a copper coil wound round a light non-magnetic cylinder, sometimes called a turret, inside the magnetic field of a permanent magnet. When a current is passed through the coil it reacts to the magnetic field of the permanent magnet surrounding it and moves at right angles to the field direction. If the current is reversed, it moves in the opposite direction. In cross section, it looks like this:

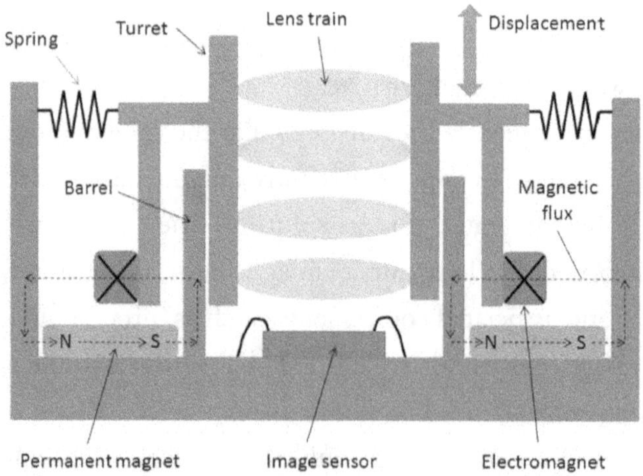

Source: *Image Sensors World*, 1 July 2011

In a phone camera, the VCM is a tiny square-shaped component that sits around the lens it controls. All the different parts – approximately ten – need to be fitted together securely to create a miniature motor. In the early days, much of this intricate and fiddly assembly work had to be done by hand. In particular, a coil of (usually) copper wire has to be wound round a coil holder to precise tolerances. Companies that emerged as suppliers to the smartphone module market were generally not those already involved in the manufacture of Voice Coil Motors in use for, say, vehicles or industrial processes but a new breed dedicated to making the more precise VCMs demanded by handsets.

The driver IC that accompanies an actuator operates at a deeper level of application than a standard ASIC, by offering the facility to embed precise instructions compatible with a particular model of phone. The software embodied in a driver IC is in the form of firmware. Firmware is data that is stored in a chip's memory and provides precise instructions on how that device should operate. In the hierarchy of software, firmware sits just above hardware. It is software embedded in hardware.

Modern electronic products, such as smartphones, receive regular software updates to their operating system. This is often in the form of what is called over-the-air firmware, remotely downloaded code that tells the hardware to behave in a new or modified way. In firmware, the algorithms are written into a Read-Only Memory (ROM). Importantly, data stored in a ROM is not lost when the device is turned off. Unlike normal software, firmware cannot be changed or deleted by the end-user. In some cases, the control software is built into the driver chip itself while in others the ASIC customer supplies the software for installation in the ROM in a nearby connected IC. Wherever it is located, the customer owns the IP embodied in the firmware.

To sum up: in the early 2000s, 1...Limited was busily engaged in gearing up to supply Helimorph autofocus actuators, together with the associated driver ICs, to the emerging camera phone market. What happened next is the subject of Chapter 8, after we have covered the history and dynamics of the new and unknown field confronting the company: smartphones.

6.

Smartphones take over from dumbphones

THE SMARTPHONE EMERGES

It is worth recalling how primitive mobile phones were prior to the first smartphone. An article in 2003 referred to the few mobiles then being made with cameras as 'picture messaging phones' and expressed the hope they would soon match the features offered by traditional cameras:

> 'Picture messaging phones are about to get a lot more intrusive. Thanks to a novel and ultra-cheap micromotor technology, cell phone cameras should soon be able to zoom and focus with the same precision as the autofocusing lenses used in expensive still cameras.' (*New Scientist*, 1 September 2003)

It was only in 2006 that the first smartphone – defined as a phone that could access the internet – entered the mainstream. This was the Nokia N95. It featured a ground-breaking 5-megapixel digital camera. The N95 significantly outsold similar offerings from its main rivals: Sony Ericsson (a Japanese–Swedish joint venture formed in 2001 to make and sell mobile phones) and the South Korean company, LG.

In 2006, aside from Nokia (the global market leader), Japan was the epicentre of the mobile phone industry, mirroring its overwhelming position in digital cameras. From then on, however, the Japanese pioneers in mobile phones rapidly lost ground. In essence, too much heed was paid to the needs of domestic customers and too little devoted to what international consumers wanted. Separately, Japanese electronics and optical companies tried to make inroads into the emerging mobile camera business but were generally not successful. The principal reason is that they stuck to the image sensor technology used for the sensors in digital cameras: the charge-coupled device (CCD). To the newcomers, it soon became evident that an alternative semiconductor technology – complementary metal-oxide semiconductor (CMOS) – was far superior for smartphone imaging purposes, as it was less expensive and consumed less power.

The arrival of the iPhone in the second half of 2007 was a truly seismic event. It wholly eclipsed the Nokia N95. What Apple had achieved was revolutionary. The iPhone combined three elements: a cyber-connected communications device, a high-resolution touchscreen, and a powerful computer capable of installing and running apps created by third-party developers. It was this third facet that was the real game-changer. Apple introduced the App Store in 2008 and encouraged those with ideas and expertise to create apps, so laying the foundation for today's multi-million strong app developer community. The response was immediate. By the end of the year, the iPhone hosted around 10,000 apps.

When the iPhone appeared, it prompted Google to rethink its strategy for Android, the mobile operating system it had acquired as an alternative to Apple's iOS, and was developing behind closed doors. Android was launched via the Taiwanese handset producer, HTC, in 2008, just ten months after the iPhone. It was offered – nominally – free of charge to smartphone manufacturers ('nominally' because it was not as simple as that: Chapter 21 explains). After a slow start, Android was firmly established by 2012 as a credible alternative to the Apple ecosystem. Non-Apple competitors scrambled to get aboard the platform. Google later consolidated its position by buying, in 2017, a 2,000-person research, development and design team from HTC for $1.1bn. It suited both parties. HTC needed the cash and Google acquired greater control over the Pixel brand Android smartphones it had launched the year before (with HTC as the handset manufacturer).

Nokia lost ground, after 2010 especially. The essential reason was software. This was a time of ferment in mobile software. Nokia's Symbian operating system (OS) was adequate until it faced competition from iOS (the iPhone's operating system) and Google's Android OS (which was based on the open-source operating system, Linux – which a few handset manufacturers used). Nokia made good hardware, but its Symbian OS was complex and cumbersome to use. It was consequently disliked by app developers. When Nokia decided to use Microsoft's Windows OS in early 2011 it sounded the death knell for Symbian. Another smartphone vendor, BlackBerry, made good hardware, but its proprietary operating system suffered from performance issues. Equally, the company was slow to create apps. Consequently, Google's Android OS gained ground fast on standard Linux, Symbian, Windows and BlackBerry.

By comparison with the N95, the 2-megapixel (MP) camera the original iPhone offered was frankly pedestrian. Clearly, the camera was not a priority at this point. This changed in June 2010 with the launch of the iPhone 4. Reviewers were full of praise for the handset's 5MP camera, considering it to have reached the point at which it was an adequate substitute for the basic point-and-shoot traditional camera. The phone also offered – for the first time – a front-facing camera.

This turned out to be a seminal moment for smartphone photography as, four months after the emergence of the iPhone 4, Instagram launched its ground-breaking app, which worked well with the improved camera. Since then, camera innovation has been heavily influenced by social media's requirement for higher quality imaging. As an app focusing exclusively on photos and video sharing, Instagram launched on mobile devices only – there was no Web version. Android users had to wait until April 2012 for the app to reach them. When it did, pent-up demand led to more than a million downloads on the first day. Subsequently, in 2013, the iPhone 5's software was directly influenced by Instagram. For the first time, it allowed users to take photos in Instagram's iconic square format. Instagram could not possibly have taken off in the way that it did without the sheer ease of use on a mobile device anywhere, anytime versus a laptop or PC. Similarly, the iPhone 4's front-facing camera allied to Apple's novel FaceTime software gave video chat a big boost.

How is it that we ended up with a mobile operating system duopoly? Symbian was never really a runner. Linux was unable to gain traction and BlackBerry's software had too many weaknesses. But surely Windows, which was well established and had a lot going for it, could have provided effective competition for iOS and Android? Windows phones were acknowledged to have good hardware, but Microsoft was unable to establish a foothold, even after it had acquired control of Nokia. By the end of 2017, Windows Mobile OS had been abandoned. The essential reason for this failure lay not in the software – which performed reasonably well – but in the platform's chronic failure to attract app developers. Microsoft was further handicapped by Google's refusal to allow Windows phones to run YouTube, one of the most popular apps. It is a pity that Windows never made the grade as a mobile operating system. A third globally acceptable OS would have been a valuable addition to competition and consumer choice.

As the graph below shows, the Android OS is installed on three-quarters of all phones sold, with iOS accounting for almost all the remainder. Of course, as Apple consciously positions itself at the premium end of the market, iOS represents a much higher proportion of monetary value. Huawei's HarmonyOS is virtually unknown in Western markets as it is effectively confined to China (for reasons discussed in subsequent chapters).

Global smartphone sales share by operating system

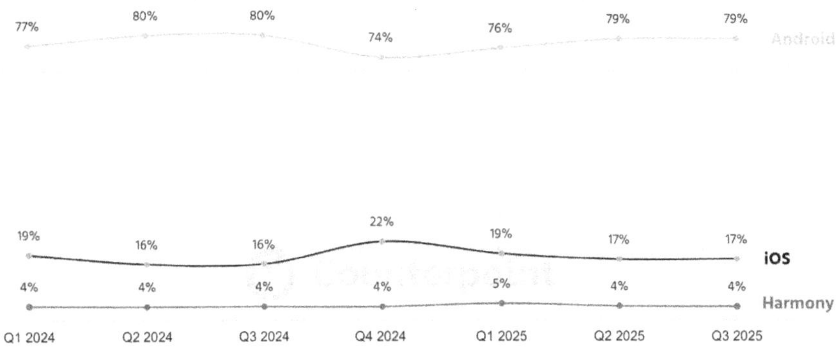

Source: Counterpoint Research

Although the iPhone camera saw significant improvement with each new generation of handset, it took Apple until 2015 to realise that a top-notch camera was an essential component in its smartphone offering. Surveys suggest that second to its essential communications and internet capabilities, the ability to take photos and videos, for social and business purposes, is a phone's most important function. We use it constantly. Consider, for example, the proliferation of QR codes since the pandemic. Android vendors flag the camera as a differentiating factor.

The evolution of the iPhone illustrates much about the way Apple does business. In 2004 the company was enjoying runaway success with the iPod – it accounted for 45 per cent of group sales. However, Steve Jobs realised that mobile phones, with their growing ability to store and play music, were a serious threat. Samsung, in particular, was making a pitch for mobile supremacy using music to drive sales. A contemporary magazine article explained:

> 'Since last summer, Samsung has introduced more than 20 phones that double as MP3 players. Of course, most of these handsets still include cameras, but music is the hot trend.' ('Samsung is Putting Songs in Its Heart', BusinessWeek, 25 April 2004)

Steve Jobs therefore did a deal with Motorola to produce a phone that could play songs from iTunes. Motorola was then a leading phone maker, along with Nokia, Sony Ericsson, Samsung, Siemens, NEC of Japan and LG, the Korean conglomerate. Why Motorola? The company could justifiably lay claim to being the originator of the mobile phone. It launched the first mobile phone in 1983, priced at $4,000 and weighing over a kilo. Here is a picture of the inventor, Martin Cooper, holding his creation, popularly known as 'the brick'.

Source: Rico Shen, 5 June 2007

Apple's arrangement with Motorola turned out to be an unmitigated disaster. The resultant phone proved deficient in every respect. It suffered from poor design, a cheap plastic case and a 100-song limit, compared with the 1,000-song capacity of the iPod. Awkwardly, users had to buy music on iTunes and then transfer it to the phone via a cable. More broadly, Motorola's mobile phone line-up in 2005 was regarded as unimaginative and it quickly lost ground to competitors, culminating eventually in the sale of the division to the Chinese company, Lenovo.

By September 2006 Jobs had lost patience with Motorola and decided that Apple should never again rely on others for the design of its products, an ethos that is central to Apple's modus operandi. The company then embarked on a project to produce a phone of its own, which it did in nine months. The table below shows the impact of the iPhone three years after its launch.

COMPANY	2010 GLOBAL SMARTPHONE MARKET SHARE (HANDSET UNITS)
Nokia	34%
Apple	16%
RIM (BlackBerry)	16%
HTC	8%
Samsung	8%
Motorola	5%
Sony Ericsson	3%
LG	2%
Palm	2%
Others	6%
Total	**100%**

Source: An amalgam of technology consulting firm estimates

Note that these figures are for smartphones only. Over 80 per cent of the mobiles manufactured in 2010 were handsets that could make calls but could not access the internet. These were soon redesignated by the industry as 'feature' phones. This is nonsensical nomenclature. Smartphones are stacked with features. So-called 'feature' phones have barely any. Surely 'dumbphones' makes more sense. It was a feeble, and futile, attempt to maintain their relevance. This book takes a firm stand on the issue: 'feature' phones will henceforth be referred to as 'dumbphones'. Dumbphones are still being produced and sold, predominately in Africa, the Middle East and India, though they represent a small and declining part of the global mobile market, certainly by value.

By 2013 the competitive landscape of the smartphone industry was changing fast. In Japan, where the major indigenous electronics firms had been responsible for much of the early innovation in mobiles, several competitors gave up while others clung on. Despite producing quality devices, Japanese producers were unable to gain a foothold in foreign markets. Instead, they concentrated on their profitable domestic market. Today the Japanese market is dominated by Apple, followed some way behind by Samsung. Sony, Sharp, Fujitsu, Kyocera and a few others are still in business, each producing a paltry few million handsets a year.

With its 'proper' QWERTY keyboard and its enthusiastic usage by corporate executives, BlackBerry could not conceive that a mere 'consumer' device like the iPhone could possibly appeal to its customer base. Having been on the slide in terms of sales since 2010, BlackBerry reached its nadir in August 2013. A financial crisis resulted in massive redundancies and attempt to sell the entire business. By 2016 it had withdrawn from handsets.

Globally, annual smartphone numbers continued to grow after 2013, reaching a sales peak of nearly 1.5bn in 2018. After that, the pandemic and economic pressures caused annual shipments to fall, to around 1.2bn in 2025. What is interesting is how well sales of premium phones – defined by Hong Kong-based technology market research firm, Counterpoint Research, as those with a wholesale price in excess of $600 – held up during this period. Affluent consumers proved to be less affected by macroeconomic pressures than lower income customers. The chart below demonstrates the trend.

Global smartphone wholesale ASP (in $) and premium segment shipment share

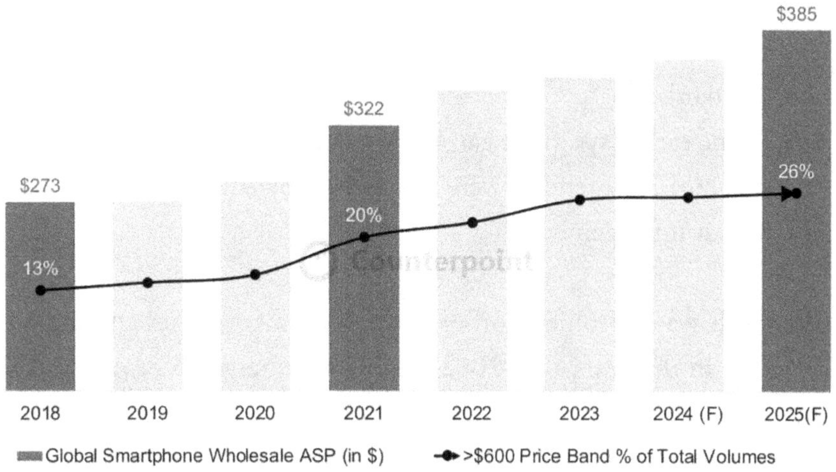

Source: Counterpoint Research

CHINA ENTERS THE FRAY

In a rapidly rising market after 2012, Chinese phone makers began to make their mark in smartphones, not only domestically but internationally. Chinese entrepreneurs pursued this burgeoning new market with vigour. There were several factors in their favour:

1. The sheer size of the domestic market. Since 2012 China has been the single largest source of global smartphone demand, accounting for about a quarter of all shipments.

2. Chinese smartphone manufacture was rooted in earlier government support for startups in the dumbphone era. *Shanzai* (knock-off) handset makers proliferated between 2008 and 2012, especially in and around Shenzhen. Their combined demand stimulated the creation of a comprehensive supply chain infrastructure. Some of these *Shanzai* developed into effective handset producers but most collapsed or remained marginal.
3. The Chinese government intervened to create favourable conditions for the industry to flourish, in the form of financial support, a benign regulatory regime and the establishment of industry standards.
4. In the early days, there can be no doubt that some of these entrepreneurial contenders were untroubled by issues of international patent infringement.

The result was a proliferation of lower cost smartphones coming out of China in the years after 2012. The rapid emergence of Android as the de facto operating system competitor to iOS provided the nascent industry with a heaven-sent opportunity, which companies seized with both hands. One estimate suggests there were 300 budding Chinese smartphone producers in 2013. Chinese companies developed the capacity to make in volume many of the mechanical and electromechanical components that go into a smartphone.

However, securing access to the necessary chips, especially the SoC, was a problem. Chinese companies lagged in semiconductor design skills as well as fabrication. Semiconductor designs suitable for smartphones, heavily protected by patents, were available on the open market from two firms: US-based Qualcomm and MediaTek of Taiwan. Patents play an important role in the smartphone business. From application to grant – assuming it is accepted as a genuinely novel contribution – normally takes a minimum of three years in the UK and two years in the

USA. Provided the annual renewal fees are paid, a granted patent gives a 20-year monopoly to the inventor (or, more commonly, the company that employed the inventor) from the date of filing.

Qualcomm especially was quick off the mark to offer its proprietary SoC designs to Chinese smartphone producers, challenged by MediaTek. The few global smartphone vendors possessing the necessary resources, notably Samsung and Huawei, chose, like Apple, to design their SoC and its associated chips internally. Qualcomm – which maintains a low public profile – occupies a pivotal position in the Android ecosystem, licensing its intellectual property to would-be handset makers. Specifically, it controls important patents related to the communications and processor functions of a smartphone and sells chips that perform these functions. With MediaTek, it remains an important semiconductor supplier to Chinese firms (a point of vulnerability for the Chinese handset industry). Both Qualcomm and MediaTek are fabless semiconductor firms, outsourcing production to foundries, principally TSMC.

In 2012 Samsung was the market leader in China, with an 18 per cent share. By 2015, however, a raft of newly formed Chinese companies had climbed aboard the Android express and were competing vigorously. Samsung's pole position quickly eroded. Huawei, an established dumbphone producer, was quick to adopt Android and was soon selling a smartphone. Another company early into the market was Xiaomi, which launched its first model in 2014. It was closely followed by Vivo and a host of others, including Lenovo, Meizu and Oppo. All of these aimed to match the features offered by the iPhone but at a substantially lower price.

Chinese firms did not just seek to copy Apple. The leading producers competed by offering additional, innovative features, including improved camera capabilities. Many of these aspirant companies failed after a few years or ended up hanging onto tiny market shares. On a much larger scale than in Japan, the huge Chinese domestic market is able to sustain a long tail of small and obscure handset producers.

In sum, from 2015 onwards, Chinese smartphone firms have had a distinct impact on world markets, by offering highly capable products at attractive prices. Understandably, the primary selling thrust was towards the rest of Asia, but inroads have also been made into Western Europe. Interestingly, Chinese firms have had very little success in penetrating the US market, with the notable exception of Lenovo selling under the Motorola brand.

Why have Chinese companies been unable to penetrate the USA? In 2010 the American market was fragmented but, by 2012, it had effectively become a duopoly, with Apple and Samsung between them commanding a 60 per cent share. The duopoly persists to this day, with the iPhone accounting for over half the market and Samsung another quarter. Geopolitical considerations were not a factor in 2012 – there was no significant political animus against China until the middle of the decade – so the minimal penetration by Chinese brands cannot be laid at the door of external events.

One reason for the inability of Chinese firms to secure market share is that America is a premium phone market, with many consumers able and willing to pay for an – expensive and status-conferring – iPhone. A further important reason is the grip on the market exerted by the three network carriers: Verizon, AT&T and T-Mobile. In theory, unified telecom standards mean that any phone will function domestically. In reality, without the support of a carrier, there is no guarantee it will work. Additionally, all new phones are confronted by a series of obstacles before becoming accepted by the carriers, including a rigorous certification process and technical compatibility tests.

Consequently, few non-Apple products (other than Samsung) were – and are – promoted by the major US networks. It is tempting to believe that, in the early days before the USA came into direct confrontation with China, there was an element of protecting the home-grown champion from Chinese encroachment. Today, keeping Chinese phones out of the US market as anything other than marginal competitors is a more-or-less overt policy. As a result, American consumers are denied the opportunity to experience models embodying innovative concepts. Any other major US industry sector with this degree of concentration and entry restriction would surely have attracted the attention of the anti-trust authorities a long time ago. Not until March 2024 did the Department of Justice launch a lawsuit against Apple, alleging the use of anti-competitive practices to maintain its monopoly of the US smartphone market.

7.

Smartphone dynamics: how it happens

MARKET DYNAMICS

Churning out 1.2 billion handsets a year, smartphone production is a vast, complex enterprise involving hundreds of companies in many different countries. A high-spec handset contains 1,500 individual components. Subassemblies, of which the camera module (or modules, as modern phones have several cameras) is one of the more complex, are delivered to assembly lines located mostly in Asia. The bulk of final assembly is located in China, with Vietnam and India steadily increasing their share. Assembly is often carried out not by the smartphone makers themselves but by specialised contract electronic assembly firms on their behalf. The global share of the leading smartphone vendors in 2025 is shown opposite.

Major rivals to Samsung and Apple that existed a decade and a half earlier are no longer to be seen. BlackBerry is still in business but as a supplier of cybersecurity software and services. Palm has long since disappeared. In 2016, HMD Global, a company founded by former Nokia employees, acquired the rights to the name and produces HMD-labelled smartphones running on Android. It also sells Nokia-branded dumbphones. Taiwan's HTC is now a minor player as is

COMPANY	2025 GLOBAL SMARTPHONE MARKET SHARE (HANDSET UNITS)
Apple	19%
Samsung	19%
Xiaomi	13%
Transsion	8%
Oppo	8%
Vivo	8%
Honor	6%
Lenovo	5%
Huawei	4%
Others	10%
Total	**100%**

Source: Omdia, a global technology market analysis firm

Sony (the alliance with Ericsson was dissolved in 2012). And, due to consistent losses incurred over the previous six years, LG announced its withdrawal from smartphones in 2021.

Not only does this ranking look very different from 15 years ago but there is a considerable change from the picture just five years previously. In the middle of 2020 Huawei briefly commanded a 20 per cent worldwide share of handset shipments. As a direct consequence of geopolitics – the US technological embargo aimed at Huawei – it is now down to a 4 per cent share, despite offering well-regarded and technologically advanced phones. The gap that opened up has been very largely filled by other Chinese manufacturers, notably Xiaomi and Oppo but also several more brands that are little known outside Asia and Africa.

Tecno is one such brand. It is invisible in the Western world. The brand is owned by a Shanghai-listed, Chinese company, Transsion, which also controls the sub-brands, iTel and Infinix. Transsion is a definite oddball in the pantheon of Chinese handset companies. Strangely, it does not sell its products in its home market, or in Europe or North America. Founded by a Chinese businessman in Nigeria in 2008, Transsion rose to become fourth equal global handset manufacturer in 2025.

Sensitivity to local requirements has allowed Transsion to pursue a successful strategy of penetrating African and Middle Eastern markets. It is the dominant smartphone vendor in Africa, with a more than one-third market share, and is now extending its operations to other lower income nations. In Africa, local customisation included cameras designed explicitly to take account of darker skin and multiple slots for SIM cards because Africans frequently switch providers in search of a better service. Similarly, when Transsion launched in India, it recognised Indians' propensity for using their hands to eat and offered fingerprint sensors that functioned with oily fingers.

As well as promoting their products worldwide, Chinese companies, led by Vivo, Xiaomi, Oppo and Realme, a spin-off from Oppo, have concentrated their efforts on India. Why India? Because the opportunity is very large. With an equivalent population to China, it is the second largest global market and is well behind China on the smartphone consumption curve. Inexpensive dumbphones are still used by a third of the population. Several home-grown Indian brands serve the dumbphone market but have had little success in smartphones.

Despite friction at the political level, the Indian authorities have encouraged Chinese – and other – smartphone manufacturers to set up assembly in India. In a bid to build the necessary infrastructure, in 2021 the government launched the India Semiconductor Mission, with the object of positioning India as a global hub for electronics manufacturing and design. There has been solid progress. By 2024

India had become self-reliant in smartphone production, with a significant export component, to the USA especially.

Seemingly, the desire to provide the Indian population with easy access to the entry-level phones made by Chinese vendors has overcome political concerns. However, tensions remain. The Indian authorities resent the 60 per cent market share that Chinese firms hold but can do little about it other than make life difficult for them ... which they have done.

PRODUCTION DYNAMICS

In terms of putting together something as complicated as a phone, the smartphone vendor is in the driving seat. The design and construction of every component is ultimately specified and dictated by them. In some cases, it is a joint effort between the phone producer and the component maker but mostly what the smartphone maker wants is transmitted from the top through the sub-assembly constructor down to the lowliest component supplier. Smartphone brands vary in the extent to which they control their supply chain. Some specify everything, while others are prepared to adopt off-the-shelf solutions.

The overriding objective of a component or subsystem maker in the electronics sphere is to secure a 'design win' in a piece of equipment. A design win is critical. Once a component is embedded in a phone, it cannot easily be dislodged. The structure used by smartphone manufacturers to create a new model handset is, of necessity, hierarchical but it is a structure in which information and discussion flow freely up and down the chain. The process begins with the provision of samples to prospective customers, who then evaluate them and decide whether the component will suit their requirements. Once contracts have been awarded, and the design parameters fixed, component suppliers are required to work closely with sub-assembly makers who, in turn, keep the smartphone producer informed. Sub-system suppliers have a say in what is specified but the ultimate decision lies with the phone manufacturer.

One attraction for the component maker specific to the world of mobiles is that phone producers constantly introduce new models, every year or even more frequently. Each new model represents an opportunity for a newcomer to persuade the phone maker to incorporate its technology: in electronics industry parlance, to get their component 'designed in'. For a smartphone, from design win to product introduction typically takes 12–15 months. The nimbler Chinese smartphone producers are certainly able to turn a device round in this timescale. Less so the two leaders, both imbued with a structured planning ethos. Samsung typically takes a few months longer to translate an initial design into a commercial product and Apple longer still (two to three years). Smartphones are a fast-moving industry but those who develop and make them are intrinsically conservative. After all, one faulty component can lead to embarrassing recalls and a huge amount of reputational damage (exemplified by the event that rocked Samsung in 2016: see Chapter 17).

Every type of electronic equipment has a selling – and therefore production – life cycle before and after launch. For mobiles, the production life cycle is short. Component suppliers will typically see large orders three months ahead of launch, followed by a ramp-up of orders for up to a year, then tailing off rapidly as a new generation model hits the market. Other electronic products typically have longer production life cycles. Take power supplies, for example. Every item of equipment that is connected to mains electricity requires a power supply. The production life cycle for a power supply is seven or eight years.

Beyond a design win, there is another issue of equal importance. Design engineers understandably get excited when they have created a working model of a device and are able to fabricate half a dozen on the bench. The next stage, out of the laboratory, is a small-scale assembly line to produce a few hundred or maybe a few thousand samples for evaluation by potential customers. A fully operating device is fine but what customers want is a production item. A working device is of no use if it cannot be reliably manufactured in high volume at a consistent level of quality and cost. In the electronics industry, this is often referred to as 'manufacturability'. Smartphone customers need to be assured that, if they design a component into their next model, it can be delivered in the millions each and every month.

An additional factor the smartphone producer has to take into account is patents. Smartphones rely on international telecommunications standards to allow them to connect to wireless networks. Organisations engaged in standard setting have developed rules and practices to ensure the efficient licensing of patents. These are known as Standard Essential Patents, or SEPs. Under an SEP regime, patent-holders are obliged to license their product to anyone at a 'fair, reasonable and non-discriminatory rate'. Several companies – Qualcomm, Ericsson, Nokia, Samsung, Huawei and a dozen others – hold essential patents, which they license to phone makers.

Mobile phones are full of components and technologies that were developed and patented by different companies. Smartphone makers are obliged to pay royalties to the owners of the IP. From time to time, companies take legal action to enforce these payments. In 2024 Qualcomm sued Transsion for violating its IP, as did Huawei in 2025. All in all, payments for patents represent a significant intangible expense, amounting to perhaps 5 per cent of the cost of an average handset.

Qualcomm is, without doubt, the doyen of the smartphone intellectual property world. The firm has an unrivalled portfolio of significant patents relating to mobile devices, with a particular strength in wireless modems, a key semiconductor component in a smartphone (explained in a later section of this chapter). Financially, it has proved to be a highly successful formula, which, in the year to September 2025, produced $44bn in revenue accompanied by a 28 per cent pre-tax profit margin.

Unlike, say, injection moulding, where plastic pellets go in at one end and a finished product emerges at the other, smartphone manufacture is an assembly job – putting together a finished product from bought-in parts. Handset makers differ radically in terms of the degree of vertical integration they embrace. Some, Apple being the exemplar, rely entirely on contracting with third parties to produce and assemble the parts they have designed and tightly specified. Others, notably Samsung, are at the other end of the spectrum, doing as much as they can to research, develop and manufacture within the corporation or its affiliates. Both philosophies are valid. In principle, Samsung has more control, but Apple has a well-honed system that expects and demands much from its more than 200 suppliers. Other manufacturers operate somewhere in between, leaning towards the Apple model as few companies have the deep and extensive capabilities of Samsung.

If there is a single location that epitomises China's domination of smartphone production it is Shenzhen, adjacent to Hong Kong, the country's nearest equivalent to Silicon Valley. Like the Bay Area in northern California, it is a hotbed of entrepreneurial activity, with new tech companies being formed the whole time. Shenzhen has grown to its current size from a fishing village in the 1980s, just as the San Francisco Bay Area was covered by fruit farms until the 1950s. It is now a 17mn population megalopolis.

Final assembly, inspection and testing of handsets is often carried out by external entities who specialise in this. These are generically referred to as Electronic Manufacturing Services (EMS) companies. EMS is a well-established sub-sector in the electronics industry. Contract assembly takes place in Asia, predominately China. Key names here are Foxconn (the major subsidiary of the quoted Taiwanese company, Hon Hai Precision Industry) and Pegatron (also Taiwanese) and Chinese-owned Luxshare Precision. Most handset makers of any size, other than Apple, use a combination of in-house facilities and outside contractors for final assembly.

Employing an external organisation to assemble a handset makes sense. The tedious, repetitive task of carefully inserting parts into a small container as it passes along an assembly line is far removed from the skillset needed to research, develop and design an attractive and functioning smartphone at an acceptable cost. A decade ago, smartphone assembly was highly labour-intensive, employing mostly young women who had migrated within China to find work. In recent years, there has been a push to automate many of the steps involved in the assembly of a handset, partly due to the increasing shortage of labour from the Chinese countryside and consequent wage increases.

Samsung, for its higher end Galaxy range, has gone furthest down this route, with a dedicated facility in Korea almost entirely 'staffed' by robots. Nevertheless, for the most part, handset assembly remains labour-intensive. Around half of all Samsung phones are assembled by Samsung in Vietnam. Chinese EMS companies are responsible for the budget-level phones that constitute a quarter of Samsung's output. Indian EMS companies account for a further 15 per cent.

In common with its fellow EMS providers, Foxconn provides an assembly service for a range of consumer electronics companies, from Amazon to Nintendo. It is, without doubt, the pre-eminent EMS supplier. The company has moved beyond offering low-margin contract assembly services into the manufacture of components and sub-assemblies and also into new sectors entirely: cloud computing, automation, robotics and electric vehicles. Foxconn has also successfully climbed onto the AI bandwagon by becoming a major factor in advanced packaging. The firm has a strong relationship with the leading AI proponent, Nvidia, integrating its chips into servers for data-centre customers, holding a 40 per cent share of this business. Assembling these servers requires competence in interconnection, power supplies, heat dissipation systems and testing. In technological terms Foxconn is highly capable, with over 100,000 active patents.

Zhengzhou, a town in central China, is the site of the largest iPhone assembly plant. It is operated by Foxconn. Pegatron assembles iPhones in Shanghai while Luxshare Precision Industry operates an assembly plant for Apple in Kunshan, a small city near Shanghai. Foxconn accounts for around half of iPhone shipments. At its peak, it employed 350,000 workers and assembled an astonishing 500,000 handsets a day. Worker numbers have declined to around 200,000 at the busiest time of the year, due to lower volumes and automation. Luxshare has successfully built a broad relationship with Apple in recent years as it seeks to reduce its dependence on Foxconn, which has historically accounted for the bulk of iPhone assembly. Apple has invested significantly in Luxshare to bring the company up to Foxconn standards. The closeness of the relationship is confirmed by the disclosure in the 2024 Luxshare annual report that 'Customer 1' (Apple) accounted for 70 per cent of turnover in that year.

Recent years have seen handset assemblers move their operations from China to places like Vietnam and India with access to large pools of agrarian labour. Geopolitical friction between the USA and China has accelerated this trend. Nevertheless, China continues to retain the great bulk of this activity due to its accumulated engineering acumen and the existence of a well-established and responsive network of component suppliers. Tim Cook, the CEO of Apple, in an interview just at the point at which Apple was stepping up its push into China, succinctly summed up the country's advantages.

> 'The popular conception is that companies go to China because of the low labor cost … but the truth is that China stopped being

a low labor cost country many years ago. That is not the reason to come to China … the reason is because of the skill and the quantity of skill in one location, and the type of skill it is.' (Fortune Global Forum, Guangzhou, December 2017)

The special tamper-resistant screws that secure the iPhone are a case in point. These could be manufactured by any industrialised nation, but Chinese firms alone are capable of making them in the quantity and at the level of quality required. Similarly, Apple is reliant on China's large cohort of production engineers to create the conditions for efficient manufacture. As a consequence, the shift of iPhone production elsewhere is necessarily a gradual process.

While EMS companies, like Foxconn, are visible to the outside observer, upstream smartphone production is dependent on another group of companies that operate entirely under the radar. The design and development of the components that go into a new model is, in over 40 per cent of the handsets introduced each year, outsourced to so-called ODM (Original Design Manufacture) firms. Smaller handset producers, such as Transsion and Lenovo/Motorola, rely heavily on the external creative expertise these firms are able to bring to bear. Even major vendors use ODMs for *part* of their output to add flexibility to their in-house capability, attracted additionally by the potential for lower cost (especially important for entry-level products). When demand for phones flips up, the big smartphone producers turn to ODMs to plug the gap. Thus Samsung, Huawei, Oppo, Vivo and Xiaomi routinely use ODMs for this purpose. The ODM sector is dominated by three Chinese-owned companies: Huaqin Technology, Shanghai Longcheer and Luxshare (which bought into this business in 2025). Apple eschews the services of ODMs. Design is entirely internal.

Overall, looking at smartphone production, there is much that goes on beneath the calm surface waters of the retail brand. The industry is characterised by an ever-changing patchwork of relationships and connections. A Samsung phone, for example, could be entirely designed and produced within Samsung or it may never have seen the inside of a Samsung facility. Not that the buyer cares. The brand is the guarantee.

Where are smartphones actually made? This is an easy question to pose but a complicated one to answer. It is asked most often in regard to Apple, as it is the only smartphone manufacturer with no internal facilities and relies on a worldwide network of subcontractors. It is an abiding myth that most of the value of an iPhone is captured by Chinese companies because that is where final assembly generally takes place and because China is the source of the majority of the parts. True, Chinese-headquartered firms constitute the largest single grouping in the pantheon of iPhone suppliers – and the number is growing.

However, the perception that China is 'responsible' for iPhone production is compelling but misleading, for two reasons. First, value rather than numbers is a far more accurate guide to the contribution made by each country to an assembled iPhone. Second, Chinese component suppliers have added – or substituted – production capacity elsewhere in Asia. Generally, the components supplied by Chinese firms are lower cost, though this is gradually changing as Chinese firms move up the value chain. Admittedly, in terms of the *number* of components delivered, China-located companies figure prominently in all phones, supplying mostly mid- or low-tech items, such as acoustics (microphones and loudspeakers), casings, connectors, batteries and metal frames. Chinese industry has a particular strength in precision mechanical and moulded components made from metal or alloys.

The great majority of the expensive components inside a handset, especially the SoC and the display panels, are made outside China. In a high-end phone, a commercial SoC is the most expensive item, accounting for around 20–25 per cent of the total build cost. (Due to its mutually beneficial relationship with TSMC, Apple is able to make its SoCs more cheaply.) As additional functionality is built in, the cost of an SoC is rising. The display panel is equally costly. And, as of the beginning of 2024, memory chips have risen sharply in price, driven by the surging demand from AI data centres.

Flagship phones contain as many as 30 semiconductor chips in addition to the SoC, performing logic, memory, communications and other functions. The great majority of these are not leading edge and only require mature process technology nodes (above 30nm) for fabrication. Semiconductors account for well over half of the value of the components in a premium phone. These chips are predominately designed by American fabless firms and fabricated by TSMC and others. Leading Chinese smartphone producers, notably Xiaomi, Oppo and Vivo, are highly dependent on these ICs – a point of vulnerability in the context of fraught Sino-American relations.

As to the geographical distribution of the value of the components that go into an iPhone, this is an exceptionally tricky calculation. There are plenty of estimates of the contribution by country to the 'bill of materials' (BoM in industry jargon) by number of firms and their location, based on the data provided by Apple each year. The price paid by Apple and the precise location of component manufacture is information that is, understandably, not in the public domain. What is clear is that the USA, South Korea, Japan and Taiwan all figure large in the value calculation.

A few firms specialise in what the industry calls a 'teardown' of electronic products to examine the nature and cost of the components. Unfortunately, these are conducted on different bases and are not all that helpful. There is a general tendency to downplay the Chinese contribution as it is much more difficult to identify. Assembly itself is not a major factor: it accounts for only 2 per cent of the total bill of materials. The problem of identifying components arises because China's extensive smartphone parts infrastructure is responsible for the manufacture of multifarious inexpensive, low and mid-tech items that are difficult to track. The following table, an amalgam of several sources, endeavours to allocate the build cost of an iPhone by geography.

Estimated 2024 value contribution to the build cost of an iPhone by country

COUNTRY	PERCENTAGE CONTRIBUTION
USA	24%
South Korea	21%
Taiwan	21%
China	20%
Japan	11%
Others	3%
Total	**100%**

Source: Author's estimates

All in all, at a guess, the proportion of the iPhone's value attributable to activities that take place in China lies in the 20 per cent range.

The second reason why the number of Chinese firms overstates the China-located contribution to iPhone manufacture is that, in response to Apple's express wish for geographical diversification, many of its established Chinese suppliers have set up manufacture in other parts of Asia. In going down this route, Apple (and others) are pursuing a back-up strategy, dubbed China Plus One. The Plus One has tended to be either Vietnam or India. In India, Chinese smartphone manufacturers have generally opted to outsource assembly to local EMS companies. Apple has invested in both of these countries by encouraging existing suppliers to set up facilities there. The company started assembling the iPhone 15 in India in 2023, in a plant run by Foxconn.

The Indian government has the objective of positioning the country as a fully fledged second source of smartphones. This is an uphill task as, currently, China alone can produce many of the mechanical or electro-mechanical parts in a handset in a cost-effective manner. To upgrade the country's capability, the government has launched the Electronics Component Manufacturing Scheme. On semiconductors, India aims to create a hub with its attendant inputs, building on the country's excellence in chip design and software. International collaboration is at the core of this ambition, exemplified by the Japan–India Semiconductor Supply Chain Partnership signed in 2023.

Apple's move is part of general Western corporate tendency to 'derisk' Chinese-anchored supply chains. Trade tensions and other sources of disruption since 2018 have highlighted the potential negatives of relying to such an extent on a single country, especially when that country is intent on flexing its military, economic and political muscles. Equally, the slowing of the Chinese economy since 2022 has encouraged local companies to look abroad.

SOFTWARE DYNAMICS

The operating system experience offered by Apple or Android is not the same. Apple's iOS is a closed system whereas Google's Android is an open system. A closed system is one that does not reveal the source code of the OS while an open operating system is public and available for anyone to use. The essential comparison looks like this:

» **Apple**. The authority Apple exercises over the iPhone app ecosystem mirrors the rigour it applies to the management of its supply chain. Apps submitted to the App Store have to meet certain minimum standards. iOS appeals to those for whom security and privacy are priority issues. No other source of apps can be accessed through the Store. However, as of 2024, under regulatory pressure, Apple was having to accept 'encroachment' by non-approved apps. In mid-2025 the Apple ecosystem hosted roughly two million apps.

» **Android**. Android is the dominant platform in most countries. The main exceptions are the United States and Japan. Most Android apps are distributed through the official Google Play Store (also called Google Play) which comes pre-installed on Android devices. Control is exercised by the insistence that Google Play is only available on devices that are certified by the Android Compatibility Program. In this way, Google continues to exert its influence over Android, ensuring advantages for its own suite of apps. Some Android smartphone vendors host their own app store, either in addition to it or as a substitute for Google Play. Examples are the Samsung Galaxy Store, Xiaomi GetApps or Huawei's AppGallery. These come preloaded on the device.

  On top of this, the Android platform provides third parties with the opportunity to offer bundled apps in their own store. There are hundreds of these third-party stores. One of the most prominent is the Amazon Appstore. Similarly, a host of Chinese technology companies offer their own app stores, the leader being Tencent MyApp. Following the Chinese government's ban on Google Play in 2011, Android in China has evolved into a distinct and separate ecosystem. All in all, Android users are able to choose from an estimated four million apps.

The purest, most basic version of Android is known as AOSP, which refers to the Android Open-Source Project. Android phones receive an operating system update each year from Google. However, these releases apply to the proprietary version of Android developed by Google – a step above AOSP – that includes Google's suite of mobile applications. As the central custodian of the Android ecosystem, Google is the only organisation that can ensure interoperability between brands.

For years, Google adopted the practice of giving its annual Android software release the name of a dessert, instead of just referring to it by number. So we have had, for example, Ice Cream Sandwich and Marshmallow. Then suddenly in 2019 there was no name attached to Android 10, just a number. Cue consternation in the Android camp! Google backed down at the end of 2021 and reverted to dessert names. Android 14 was called Upside Down Cake. Android 16 is Baklava.

Android phones tend to come with a large number of pre-installed apps and services. The majority of these are useful and valuable. However, some of what is pre-installed is bloatware – apps that send spam notifications or run unnecessarily in the background. This is because Android vendors often partner with developers or commercial companies and pre-install apps and games from them. Chinese vendors of entry-level phones commonly use bloatware to keep the price down. Commissions earned from third parties offset the low or even non-existent margins made on these handsets. Some of these apps can be removed by the user or disabled, but not all.

CHIP DYNAMICS

The mobile phone industry is currently the largest single source of demand globally for semiconductors. According to Counterpoint Research, chips for smartphones accounted for 23 per cent of the total semiconductor market in 2024. However, as shown in the graph below, other industry sectors, notably servers for data centres and automotive, are predicted to grow faster.

The operating system inside a phone has a direct relationship with the integrated circuitry inside that phone – and with the SoC in

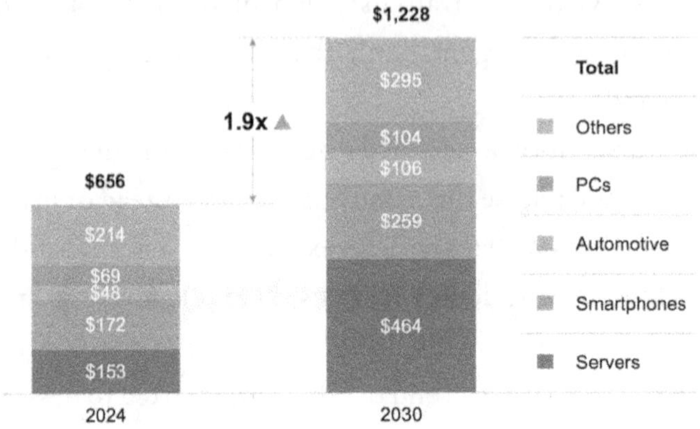

Source: Counterpoint Research

particular. As we have seen, Apple often designs its own chips for the iPhone, its self-designed SoC being the most important of these. The Android universe, on the other hand, sources its SoCs from a handful of merchant semiconductor providers, led by Qualcomm and MediaTek, marketing their own version of the System-on-a-Chip. Phone manufacturers have an understandable desire to control their destiny to the greatest possible extent by setting up an internal chip design capability, with the ultimate goal of a self-designed SoC. This is, however, a major resource commitment with numerous hurdles along the way.

What exactly is a System-on-a-Chip? (In passing, we should note that it goes by several names: chipset, Application Processor or AP, or even just processor. In this book we stick to SoC.) An SoC consists of an integrated circuit that integrates all or most of the essential elements of a computer or other electronic system. It usually contains digital, analog and mixed-signal functions. The IC commonly includes a Central Processing Unit (CPU), a Graphics Processing Unit (GPU) and a wireless modem (modulator/demodulator) for the communication function, alongside other components performing additional functions – all on a single piece of 'silicon real estate' (as industry participants are wont to call it).

The SoC truly functions as the 'brain' of a handset. The idea of miniaturising several functions on a single semiconductor chip dates back to the 1970s but received a boost from the mobile revolution of the 1990s. SoCs are especially attractive to the portable equipment designer because they offer low power consumption. Nowadays, SoC technology is extensively used in a wide range of applications, from automotive systems to domestic appliances. A major growth area is data centres, the backbone of the internet.

The idea of combining several electronic components on a single silicon chip began with the invention and commercialisation of the microprocessor by Intel in the early 1970s. A microprocessor is what we would now call a Central Processing Unit, or CPU. Microprocessors lay at the heart of early personal computers. At around the same time, Texas Instruments was working on a similar concept and came up with a comparable multi-functional integrated circuit for a different, less demanding, market segment – initially calculators – which it called a microcontroller. A microcontroller is an inexpensive SoC, easy to program, and dedicated to performing a particular task. Microcontrollers are ubiquitous in modern life and continue to proliferate, in cars, microwave ovens, washing machines, programmable thermostats, electric toothbrushes and many more.

The System-on-a-Chip was the logical step up from the microprocessor for an application requiring small size, high performance and power efficiency: in short, a smartphone. Consequently, it was the smartphone that drove the development of the SoC. In the creation of an SoC, Apple has the advantage that its design – being in-house – can be solely focused on optimising the chip for the iPhone model currently on the drawing board. The company's early SoCs were fabricated by Samsung but, since 2014, Apple has used TSMC. 'Apple Silicon' is the all-encompassing term used to describe semiconductor components developed by the company.

For Apple, there is no compatibility problem. Not so for Android SoC purchasers buying a standard design from the likes of Qualcomm or MediaTek. The Android customer's SoC is never going to be wholly aligned with the hardware design. SoCs can, however, be customised via software to try and achieve the specifications that the customer is aiming for. As a maker of both phones and semiconductors, Samsung has a particular problem. The question it has to ask itself is: should the SoC design it produces prioritise the needs of its in-house mobile phone business in the hope of giving it an edge, or should it focus on the requirements of external smartphone customers? While negotiating this tricky terrain Samsung has frequently encountered difficulties (as we shall see later).

Visually, an SoC has little to commend it. As seen here, it looks like just like any another microchip. Snapdragon is the brand name used by Qualcomm.

Source: Pocket-Lint, 16 November 2022

Qualcomm and MediaTek vie for leadership in SoCs, with approximately one-third of shipments each. However, in terms of value, Qualcomm is well ahead as it controls a high share of the premium phone market, with a particular focus on SoCs for 5G-enabled devices. In 2025 MediaTek generated sales of $19bn while Qualcomm reported SoC revenues of $28bn from handset customers. Admittedly a crude comparison, it makes the point about Qualcomm's higher average price. MediaTek is challenging Qualcomm at the higher end of the scale.

Apple, Qualcomm and MediaTek are fabless semiconductor companies: the manufacture of the SoCs they design is outsourced to a semiconductor foundry, predominately TSMC. Irritatingly, the media has a habit of referring to Apple, Qualcomm and MediaTek (and others, such as Nvidea) as chip *makers*. This is a misnomer. They are chip *designers*, but it may serve their purpose to leave the public with the impression that they fabricate their own chips.

Qualcomm SoCs power many Android phones, but the company hides its light under a bushel. Qualcomm could have adopted the Intel approach to personal computers and mounted a long-running advertising campaign to make the purchaser aware of 'Snapdragon inside' but chose not to. However, this policy is changing. In September 2023 Manchester United football club confirmed an expanded partnership with Qualcomm. Snapdragon is now the club's front-of-shirt sponsor.

How did Qualcomm come to occupy such a pivotal position in the smartphone firmament? The company began life in 1985, at a time when wireless communications relied on analog technology. Irwin M. Jacobs and six fellow engineers from MIT (Massachusetts Institute of Technology) came up with a digital solution. Qualcomm pioneered a technology called code division multiple access (CDMA) that allowed multiple cellular users to share spectrum bandwidth efficiently. From that time on, Qualcomm engineers have been at the forefront of generational advances in cellular communications technology, from 2G to the current 5G.

Qualcomm is known especially for its modem chips that connect phones to the mobile network. It has been assiduous about patenting its inventions and, as a result, maintains a firm grip on the communications technology embodied in every mobile phone. Different cellular networks operate on different wireless networking standards and Qualcomm enjoys a lead – some might call it a stranglehold – in chips that support these standards. It maintains a particular strength in proprietary modem chips. Any handset maker building a device that connects to the mobile network has to pay the company a licence fee. Indeed, it treats these patents as a gateway, pursuing a strategy of 'no licence, no chips'. Customers are not allowed to buy cellular chips unless they first sign a licence for the patent portfolio. Even Apple has to bow to the strength of Qualcomm's IP in this area. Anti-trust suits have failed to force Qualcomm to change its policy.

Structurally, Qualcomm is a strange corporate beast. It really consists of two companies in the same field pursuing quite different business models. These are:

- **Qualcomm Technology Licensing (QTL)**. QTL controls the patent licensing business, holding nearly all of Qualcomm's patents. Profit margins are extremely high because there are minimal expenses associated with this business.
- **Qualcomm CDMA Technologies (QCT)**. QCT derives revenue from the sale of integrated circuits and software that it has developed for application in mobile devices and a wide range of other electronic products. Snapdragon-branded ICs for handsets continue to account for two-thirds of sales. SoCs are sold to Samsung and to all the major Chinese phone producers. Margins are good but nowhere near those earned by QTL.

Qualcomm and MediaTek are the two powerhouses at the centre of the Android universe though few smartphone buyers know it. Qualcomm's 2024/25 annual report disclosed that just three 'customer/licensees' were responsible for 54 per cent of revenues. Chinese customers are believed to account for more than half of revenues. Apple does not buy SoCs from Qualcomm, but it does buy modem chips, despite a multi-year effort to develop its own modem IC, which eventually bore fruit in 2025. Equally, Samsung buys SoCs from Qualcomm despite possessing an internal capability.

Designers of Android SoCs like to give them user-friendly names. Qualcomm offers the Snapdragon range. Samsung has its Exynos series. Exynos is Greek for 'smart' (as in clever) and 'green' (presumably as in environmental). The mainstream MediaTek range is named Dimensity, a mishmash of an obsolete English word for dimension. Huawei has a design subsidiary, HiSilicon, which provided SoCs for about two-thirds of its internal phone requirements until American pressure on TSMC meant that it had to cease shipments to Huawei in September 2020. Huawei's SoC is called Kirin. A Kirin is a mythical East Asian creature. Anyone familiar with Japan will recognise it as a well-known brand of beer!

Huawei's travails have prompted other Chinese smartphone vendors to step up their attempts to create a home-grown SoC. Reliance on foreign designed and made SoCs is clearly a source of vulnerability for the Chinese handset industry. From a national perspective, the focus of these efforts has been on a heavily state-backed fabless semiconductor firm called Unisoc, which has made modest progress towards creating a commercially competitive product.

Some Chinese smartphone manufacturers have gone down the resource-intensive road of developing their own SoC. Oppo is a case in point. It set up a subsidiary in 2019 called Zeku Technology, based right in the heart of Silicon Valley, to design components including, in due course, a SoC. Reportedly, 3,000 engineering staff were recruited by the offer of high salaries. Then without warning, in May 2023, a slump in handset sales forced Oppo to disband the unit. Some might see this as hubris. Undeterred by Oppo's experience, Xiaomi reiterated its commitment to the dream of designing its own SoC and, in 2025, it launched its XRing O1 processor.

Google too has gone down the SoC path. With its roots in software, Google comes at the business from a different angle. The company's initial foray into hardware in 2010 was half-hearted but, in 2016, it launched its Pixel brand smartphones. Pixel was pitched as an upmarket device with a first-class camera. The superior camera experience offered by Pixel phones drew upon Google's deep software resources, making use of a technique called 'computational photography' (covered in detail in Chapter 10) to augment and improve upon what the phone's lenses capture. For Google, moving on to designing a proprietary SoC was a logical step, if only to enhance the photographic features the Pixel phone offered. Tensor, as Google's SoC is named, was developed in collaboration with Samsung and is, at root, a customised version of Samsung's Exynos SoC.

Memory chips are essential for the functioning of a smartphone. There are three types of memory used in mobile devices:

1. **Random-Access Memory (RAM)**. This stores data currently being used by the processor. It is temporary memory, and the data is lost when the device is switched off. The memory ICs used inside smartphones for this purpose are known as **DRAM** (Dynamic Random-Access Memory) chips.
2. **Read-Only Memory (ROM)**. A key feature of a ROM is that the data is retained when the device is powered off – it is non-volatile. Nor can the instructions in the memory, once written, be erased or modified. It exists to hold pre-programmed basic instructions in the form of firmware.

3. **Flash memory**. This type of ROM combines non-volatility with the ability to erase and rewrite. It is so called because the erasure process – seemingly – happens in a flash. Portable devices make extensive use of flash memory to store videos, photos and music. It is possible to erase the firmware in a flash memory ROM and reprogram it. Flash memory is low cost, compact in size and consumes little power. The trade-off is that it is relatively slow in operation.

The smartphone memory chip business is highly concentrated. It is dominated by Samsung, a second large South Korean company, SK Hynix, and Micron Technology of the USA (located – improbably – in Idaho because the original finance came from the local 'potato king', inevitably giving rise to quips about 'chips from chips'). In the DRAM sector, where SK Hynix leads, Chinese companies are beginning to make inroads.

COMPONENT DYNAMICS

Turning to the internal structure of a smartphone, a handset consists of around eight major subsystems. With the cover off, an iPhone 15 Pro looks like this.

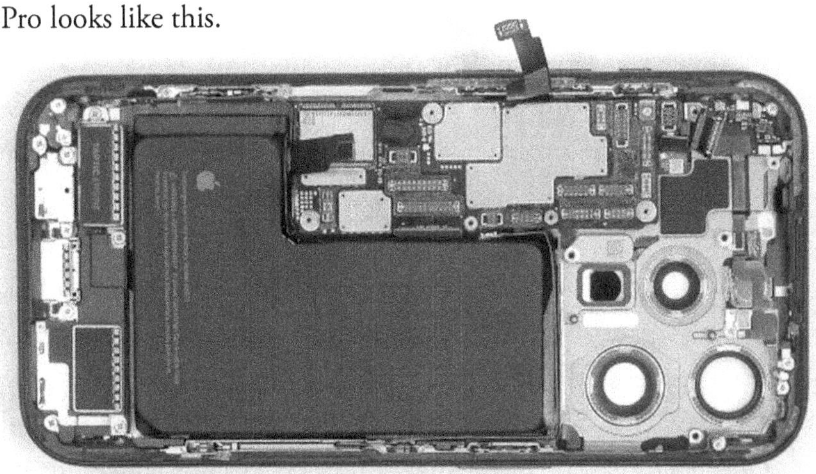

Source: The Verge, 21 September 2023

The image on the previous page is hardly illuminating. So tightly packed are the components it is hard to distinguish them (with the exception of the large black blob that is the battery). The diagram below provides a highly simplified guide to the major subsystems that together contain the components in a flagship smartphone.

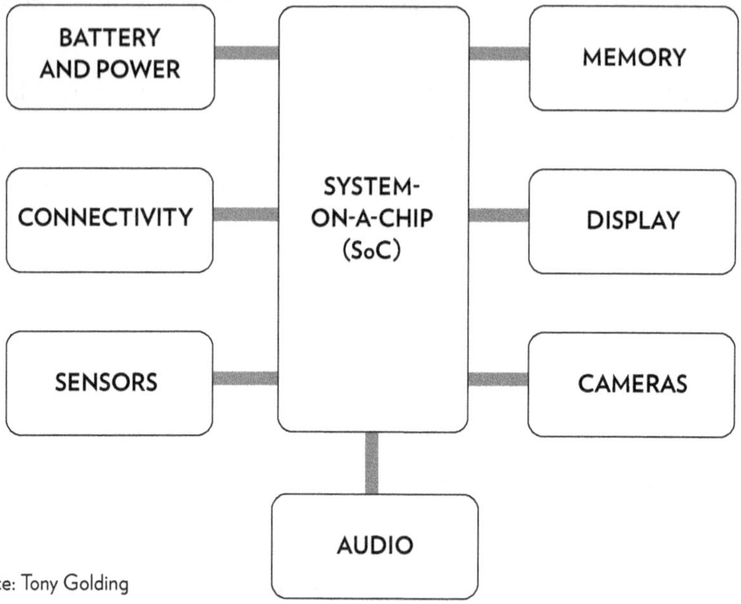

Source: Tony Golding

The spotlight is, as ever, on the SoC. Figuratively, it stands at the centre of the stage. Significant technological advances have been made since 2007 in all the subsystems – and the components within them – listed below.

- **SoC**. The SoC includes several important components, often including an on-chip modem, a trend that has gathered pace with the development of 5G.
- **Display**. Basically, smartphone displays are either LCD (Liquid Crystal Display) or, increasingly, OLED (Organic Light-Emitting

Display). Samsung is the dominant supplier, followed by the Chinese company, BOE, and LG Display of Korea.

» **Connectivity**. This refers to the various ways a device can connect to the outside world. These include cellular, Wi-Fi, Bluetooth and NFC (Near Field Communication) for Apple Pay and Google Wallet. Cellular connectivity requires two elements: a modem to convert digital data into RF (radio frequency) signals and vice versa; and RF chips to process the resulting analog signal. Qualcomm dominates the first category. The RF processing area is essentially the province of low-profile American fabless semiconductor companies like Broadcom, Qorvo and Skyworks.
» **Memory**. This subsystem includes RAM, ROM and flash memory plus the SIM card or e-SIM.
» **Cameras**. A modern phone features up to seven separate camera modules. What these are and how they fit inside a handset is explored in Chapter 20.
» **Battery and power management**. The battery is controlled and monitored by chips supplied by semiconductor firms specialising in power electronics. Once again, Qualcomm figures prominently as well as the Japanese semiconductor firm, Renesas, and Texas Instruments.
» **Sensors**. Smartphones contain as many as a dozen sensors, including an accelerometer (detects movement, such as steps), a gyroscope (detects rotation), an ambient light sensor and a fingerprint sensor (for user identification on phones without face recognition).
» **Audio**. Speakers and microphones.

8.

Enter the Helimorph

CHASING A DESIGN WIN

In 2002 1...Limited identified autofocus in camera phones – as they were then called – as the key target for the Helimorph actuator. All the evidence pointed to camera phones being at the beginning of a steep growth curve. Each new model represented an opportunity for a newcomer like 1...Limited to persuade the phone maker to incorporate its technology: to get the new component 'designed in'. In this, the company was up against the incumbent technology: the Voice Coil Motor. It soon became clear it would not be easy to shift prevailing engineering thinking. VCM had its drawbacks – and the Helimorph had advantages – but it was a tried and tested technology for moving camera lenses. Those who designed and specified mobile phones were familiar with VCM. To stick with it was the low-risk approach.

By late 2002, 1...Limited was in discussion with several companies, including Nokia and Samsung, that were actively looking to include Helimorph-based autofocus in their next generation of mobile phone cameras. Subsequently, 1...Limited was approached by two Taiwan-based camera module integrators, Premier and Lite-On Technology. Camera module integrators are companies that put together complete camera sub-systems for mobile phone manufacturers (fully explained in Chapter 10).

On the supply side, 1...Limited contracted with TDK to manufacture Helimorph actuators. Transferring the technology was not straightforward and took time. For TDK, making Helimorph actuators was a departure from their standard process and 1...Limited had to develop and build for them an automatic winding machine capable of mass production. 1...Limited also contracted with Sunnytec, a Chinese-located, Taiwanese-owned piezoceramic producer, to set up a manufacturing line, with production quantities scheduled for the second half of 2004.

At the same time, 1...Limited commissioned the development of two driver ASICs. One was from a subsidiary of Flextronics, a large contract assembler. (It changed its name from Flextronics to Flex in 2015, dropping the 'tronics' to demonstrate to customers that it offered more than Electronic Manufacturing Services.) Flextronics Semiconductor provided ASIC design services to commercial customers. The second ASIC designer was AMI Semiconductor, a small American firm.

FORWARD WITH FLEXTRONICS

Then, at the start of 2004, an event occurred that suddenly moved the project to get the Helimorph designed into a handset up a gear. Flextronics, having had a good look at the Helimorph, decided they liked it. This was a distinct breakthrough as Flextronics assembled phones for several well-known brands at that time: Sony Ericsson, Motorola and Siemens. Between them, these three controlled about a quarter of the global mobile market in 2004. At this point, Flextronics was the world's largest Electronic Manufacturing Services (EMS) company – until it was overtaken by Foxconn the following year.

Flex/Flextronics is a curious corporate animal. Originally set up in California, since 1994 it has been a US-listed company with its headquarters in Texas. Losses forced the company to pivot towards its profitable operations in Asia. Subsequently, it built a global infrastructure for high-volume manufacturing and expanded its purchasing and engineering capabilities. In the 1990s, there was a fashion for brand-based corporations to transfer their manufacturing assets to EMS companies. Sony Ericsson was one of these: it transferred all its mobile phone production facilities to Flextronics in 2001. EMS is a low-margin activity. Adding on design and component manufacture promised better returns, so Flextronics went into the business of producing parts and components, primarily for use in its own operations.

Several acquisitions were made to shore up Flextronics' component and sub-system capability. One of these was Flextronics Semiconductor. Another was the $70mn purchase of the camera module division of Agilent Technologies. Flextronics simultaneously acquired a further module assembly operation in Malaysia. These acquisitions offered the company the opportunity to insert its own module into the phones it was contracted to assemble and capture the margin that would otherwise go to outsiders.

Very early in 2004, Flextronics approached 1...Limited with a proposal to evaluate – at its own cost – Helimorph actuators in camera modules to assess their reliability and functionality in mobile phones. In parallel, Premier and Lite-On were doing the same. Then the discussions became three-way: between 1...Limited, the handset makers who had shown interest and camera module suppliers. To 1... Limited, a deal with Flextronics was clearly an attractive proposition as it provided captive access to smartphone customers as well as the all-round 'embrace' of a major corporation. Commercial negotiations with Flextronics continued until November, when the agreement was finally signed. Flextronics' purchase of the Agilent camera module facility served to confirm to 1...Limited's management that they had made the right decision.

Meanwhile, back in the world of technological reality, the project was encountering delays. The sample Helimorph actuators delivered to Flextronics failed to pass the 'drop test' that all phone manufacturers insist on, which necessitated some redesign. (As part of a suite of tests before they leave the factory, to ensure that phones can withstand real-life knocks, handsets are dropped numerous times from a height of 1.5 metres in such a way that they hit the ground at all angles.) The image below shows what drop-test equipment for portable electronic devices looks like.

By February 2005, however, things were looking up. Flextronics had successfully completed drop testing and there was interest in the generic version of the Helimorph, prosaically branded the 1AF. Better still, Flextronics had a contract to supply camera modules to Sony Ericsson for the upcoming P990 model, with projected volumes of 160,000 units a month by the final quarter of the year and the prospect of one million a month in 2006. The Flextronics camera module would replace an expensive Sony module in the current handset that consumed a lot of power.

Source: Labtone Test Equipment Co.

After years of losses, Sony Ericsson was finally in a growth phase, with rising sales and profits. In 2005 the joint venture planned to make over 50mn phones, giving it a global market share of 6 per cent. To iron out any issues before Flextronics gave the go-ahead for mass production, two pilot plants were set up, one in Cambridge. Meanwhile, Sony Ericsson was doing its own drop testing, conscious that ceramic required careful handling. The Helimorph did, however, reach the necessary threshold and, in June 2005, Flextronics ordered several million dollars' worth of equipment for mass manufacture.

SONY ERICSSON AND THE P990

All seemed to be set fair – until the bad news. In August 2005, Flextronics told Mark Shepherd, the CEO of 1...Limited, who was the main conduit for information, that the project would be subject to a three-month delay. And that the planned volume would be cut in half. Mark was unable to obtain a clear explanation for the setback. By November, though, the project appeared to be tracking the new timescale and optimism returned. Sony Ericsson had announced the P990, featuring a 2MP autofocus camera, in October, saying that shipments would start in the first quarter of 2006. The Helimorph actuators would be delivered to Flextronics by Sunnytec, not TDK as first envisaged.

Behind the scenes, pretty well all of 1...Limited's engineering resource was collaborating with Flextronics to translate the prototype and pilot line experience into a camera module capable of volume manufacture. Flextronics' skills lay in the volume insertion and connection of sourced items – not in making components or subsystems in mass production quantities. January 2006 saw the first camera modules emerge from the production line in Flextronics' facility in Zhuhai, adjacent to Shenzhen. 1...Limited engineers continued to work with Flextronics' local engineers to overcome the remaining manufacturing issues. But yields of the finished product remained half of what was needed. Then came the news in February that production of the P990 would be delayed by a further three months for software reasons. For Flextronics, this was actually a positive, as it provided breathing space to get the yield issue under control.

Then, in late March, came a bombshell. Sony Ericsson had decided to adopt an alternative AF actuator mechanism for the P990. The party line from Sony Ericsson was that they did not have enough faith in Flextronics' ability to deliver consistently in sufficient volume ahead of the launch date in May. The reaction within 1...Limited was understandably gloomy after the strenuous efforts to get the show on the road but everyone took comfort from the interest being expressed in the generic 1AF by several handset manufacturers, including Nokia. However, all these producers demanded working samples, accompanied by comprehensive drop test data, before being willing to commit themselves. In the event, the P990 did not emerge until August 2006, ten months after it was announced. The software-related delay gave Sony Ericsson time to come up with an alternative solution, most probably the expensive Sony camera module used in the previous model. In the wake of these events, Mark Shepherd left 1...Limited.

What went wrong with the project to get the Helimorph into the Sony Ericsson P990? Without access to the relevant personnel at Flextronics and Sony Ericsson it is impossible to say for certain. And what we do know is only from the 1...Limited perspective. A few points are worth making. Flextronics was almost certainly overconfident – maybe naïve – in its belief that it could rapidly ramp up a product only made in small quantities into something that could be churned out at a rate of hundreds of thousands a month. Equally, 1...Limited quite reasonably assumed that a $16bn sales corporation would have the resources to take a small-scale operation and turn it into a production-scale operation without undue difficulty. There were misconceptions on both sides.

Presumably Flextronics' self-belief derived from their experience in the mass assembly of fully fledged parts. But module manufacture was a new business for the company, with different dynamics. It is telling that Flextronics subsequently sold its camera module operation for $23mn in 2012, considerably less than Agilent's $70mn acquisition cost. Overall, it seems safe to conclude that Flextronics' eight-year excursion into camera modules was not a happy one. Flextronics today – as Flex – continues to thrive but not as a consumer electronics EMS company.

A further factor was that 1...Limited only had limited access to the end customer, Sony Ericsson, at the right level. Understandably, Flextronics was unenthusiastic about facilitating direct contact, so information tended to flow through them with whatever filters they chose to apply. To some extent, this is the fate of any small firm located in the nether regions of the smartphone supply chain. The phone vendor is aware of their existence but expects the intermediary to act as the conduit.

Quite what caused Sony Ericsson to do an abrupt *volte face* is not easy to pin down. One possibility is the difficulty the Helimorph had in passing the drop tests. Sony Ericsson must have been aware of ceramic's intrinsic lack of robustness. Was the actuator's passing margin too slight for them to be fully satisfied when they did their own set of tests?

During 2005 it became apparent to Tony Hooley and his colleagues that the PZT-based Helimorph actuator might be too fragile to provide the reliability that smartphone producers required. So the R&D team began to look for an alternative technology – a method of moving lenses that worked on different principles. As Tony puts it laconically in his Linked-In profile, 'Ceramics not being tough, disliked drop-tests, so we then developed a NiTi SMA 25μm wire motor ... as an AF actuator'. The next chapter explains what this is and what the team did with it.

9.

Figuring the way with SMA

SECURING THE WIRE

The penultimate sentence of the previous chapter refers glibly to 'NiTi SMA' as the new actuator technology that 1...Limited had invented. NiTi refers to the particular type of wire used (explained below). But what is SMA? SMA stands for shape memory alloy, an alloy that can be deformed when cold but returns to its pre-deformed (remembered) shape when heated. It is engineered during manufacture to have a default state or shape above a certain (higher) temperature – its transition temperature. Typically, the alloy is heated by passing an electric current through it.

In 1...Limited's case, the actuator mechanism takes advantage of the expansion and contraction of a particular SMA material: a special kind of very thin wire. Colloquially, it is called muscle wire, memory wire or smart wire. The input of a tiny current creates motion, which is controlled by driver ICs and software algorithms in the same way as the Helimorph. This movement is the result of a 4–5 per cent contraction in the length of the wire, a cycle that can be made to occur repeatedly and with incredible rapidity. The diagram below shows how it works.

In going down this route, 1...Limited was building on an existing foundation. The shape memory effect was first noticed in the

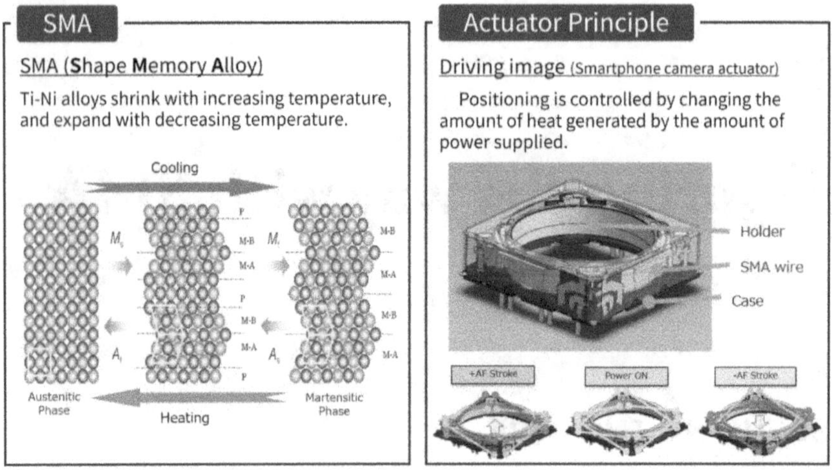

Source: Alps Alpine

1930s. Several metals in combination possess this characteristic. However, the alloy that has proved to be most valuable is nickel-titanium, which was discovered by accident in 1963 by the United States Naval Ordnance Laboratory. It was subsequently developed and commercialised under the trade name Nitinol (an acronym for Nickel Titanium Naval Ordnance Laboratory). The Nitinol market took off in the 1980s, largely due to improvements in the understanding of the material's properties. Mobile phone antennas, eyeglass frames and orthodontic arch wires made from Nitinol first saw large-scale usage during this decade.

Today, SMA wire is widely used in biomedical applications – Nitinol parts have characteristics that make them eminently suitable for implants – and in the aerospace, automotive, robotics and consumer sectors (golf clubs and underwire for brassieres). Nitinol tends to be the favoured for its unique properties. Nitinol does not have a single recipe. It can be modified by changing the proportions of nickel and titanium, or adding other elements, depending on application.

Once the concept of an SMA actuator had been satisfactorily proven in the laboratory, there was an urgent need to develop a relationship with a supplier of Nitinol wire. Worldwide, there are now a dozen well-established manufacturers. Back in 2006 it was more of a niche industry, populated by a handful of small, private companies. One of these was Dynalloy, a Californian firm founded 15 years earlier. Dynalloy manufactured Nitinol wire under the trade name Flexinol and appeared to be capable of supplying wire of the standard required. 1...Limited set up a relationship with Dynalloy via an agreement that gave 1...Limited exclusive rights to purchase the material for variable focus camera modules.

How did the notion of using SMA wire as an AF actuator mechanism originate? The idea of using Nitinol was first conceived at 1...Limited in October 2005 as researchers cast around for an alternative to piezoelectric ceramic. The team spent the next year evaluating SMA. In September 2006, the company revealed to shareholders that:

> 'In parallel with the development of (the current AF actuator), we have developed another technology to overcome the major manufacturing complication which emerged from the Flextronics/Sony/Ericsson saga. The basis of the intellectual property has been protected through patents filed.'

1...Limited filed the first patent applications for an SMA actuator in 2006. Many more were filed in the following years relating to different aspects of the functioning of the device. Each of these patents has between two and six inventor names on them, confirming that SMA was very much a team effort. Apart from Tony Hooley, the other inventors named in the applications were Ben Brown, James Howarth, David Livingstone, David Richards, Richard Topliss and Dominic Webber.

The idea of using SMA material for an actuator came from a mechanical engineer, David Livingstone. His initial concept was to use SMA to make a bending actuator (like the piezoceramic Helimorph) out of metal as this would be extremely tough and robust, unlike the earlier ceramic incarnation. This necessitated using SMA metal in sheet form, which proved to have numerous practical difficulties. The engineering team wrestled with these problems for some time until another mechanical engineer, Richard Topliss, realised that the bending principle could be dispensed with altogether, and instead the SMA material could be employed directly in the form of a thin wire. At this point, Richard had been with 1...Limited for seven years, the last two years as chief technical officer.

A major downside of applying for a patent is that subsequent publication places the invention in the public domain, with the risk that it alerts someone else to it. This is why it is important to move quickly to shore up the defences by patenting as many design features of the invention as feasible, with the object of placing numerous hurdles around the core advance in knowledge. Consequently, patent applications were not confined to the actuator's method of operation but covered other aspects, such as 1...Limited's coating on the Nitinol wire to enhance its performance.

Separately from patents – but no less important – we should not forget the protection afforded by ownership of the software algorithms in the driver IC that controls the movement of the actuator. Chapter 5 explained that the firmware embodied in a Read-Only Memory (ROM) is proprietary. These built-in instructions are not impossible to replicate but to do so would take considerable time and effort. As such, the control software is a critical element in the defensive wall around SMA.

Once 1...Limited was satisfied that an SMA-based actuator could meet reliability tests, a pilot line was set up early in 2007. With the Flextronics experience in mind, emphasis was placed from the beginning on establishing assembly processes and quality controls applicable to mass production. Keen not to promote SMA before it was ready, the sales team talked in general terms to camera module manufacturers, some of whom requested sample devices for evaluation. Handset makers – who were also contacted – were then pressing suppliers for smaller, thinner AF actuators and SMA appeared to have the potential to meet this requirement.

The next issue for 1...Limited was finding a capable manufacturing partner to license. 'Capable' because, after Flextronics, they knew what they were looking for! Discussions were held with two potential licensees: Seiko Instruments of Japan and the American conglomerate, Tyco. Part of the Seiko Group – famous for its watches – Seiko Instruments, manufactured mainly electronic components and printers. Seiko was chosen for (to quote Tony Hooley in a subsequent press release) their 'long established expertise in the manufacture of miniature mechanical assemblies'. Seiko indicated they would use production facilities in Japan and Thailand, and told 1...Limited they expected to be able to make the actuator in volume by August 2007.

THE CAMERA MODULE LABYRINTH

In February 2007, 1...Limited announced the launch of an innovative SMA-based autofocus actuator, taking miniaturisation to a new level. By May, things were looking up. Samples were out with 23 potential customers. Firm interest had been expressed by three camera module integrators: Lite-On Technology, STMicroelectronics and Samsung Electro-Mechanics (SEMCO), the component and sub-system affiliate of Samsung Electronics.

Lite-On Technology, known to 1...Limited from Helimorph days, was looking for an actuator in a camera module that it believed it was in pole position to supply to Motorola for a new handset. This was a particularly attractive prospect as Motorola was then the No. 2 smartphone vendor globally. Lite-On was the first technology company to go public on the Taiwan Stock Exchange in 1983 and has been conspicuously successful since. The Lite-On product range covered phone camera modules, LEDs, keypads and various types of sensors. As such, it represented a credible vehicle for 1...Limited in its quest to get SMA accepted by smartphone manufacturers. A licence deal with Lite-On was signed in mid-2007.

Camera module integrators were 1...Limited's immediate customers. As we noted earlier, it is they who need to be persuaded – together with the ultimate customer, the handset maker. In 2007 the leaders in the nascent camera module integration business were Sharp of Japan, Samsung and STMicroelectronics. Behind them lay a string of smaller competitors, including Lite-On. This is what the camera module industry – CM suppliers in the diagram below – looked like in 2010 according to a presentation to CML shareholders. Note the convoluted nature of the commercial relationships, as depicted by the connecting lines between the boxes.

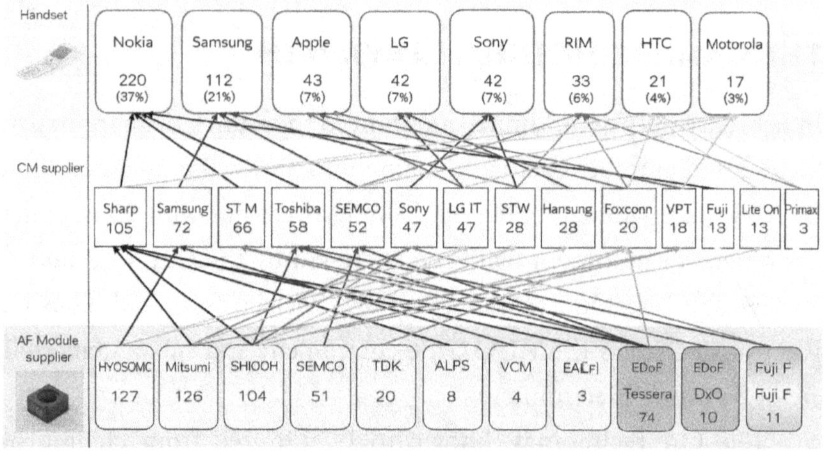

Source: Cambridge Mechatronics Ltd (CML)

The market structure of the camera module industry has changed materially since 2010. Some of the companies in the diagram above are still active but the majority have fallen by the wayside. The next chapter looks more closely at the camera module business.

The really big difference between then and now is the stellar rise of LG Innotek from obscurity in 2007 to undisputed first place today. At that stage, LG Innotek was just entering the camera module business. It is the component manufacturing subsidiary of the South Korean *chaebol*, LG. (*Chaebols* are the handful of multi-industry companies that dominate the South Korean economy, Samsung being the pre-eminent *chaebol*.) By 2010 the company, whose manufacturing relationship with Apple dates back to the late 1990s, had secured a contract for the iPhone. LG Innotek has been the principal supplier of iPhone camera modules ever since. On the back of Apple, it is now the market leader.

The early days of the smartphone camera module industry was marked by much jockeying for position. Everyone was trying to get close to the major mobile producers, in an effort to obtain preferred supplier status. The object was not just a design win for a particular model but a continuing relationship. At this point, however, mobile makers tended to run a contest for each smartphone model. The major exception was Nokia. In 2004 Nokia and STMicroelectronics got together to create a set of standards for the emerging camera mobile industry.

ST's principal business is semiconductors, including image sensors, which it already supplied to Nokia in quantity. An entry into camera modules, presumably as a means of cementing the Nokia relationship, was a sideways move for the company. It was effective: by 2006, ST was in the enviable position of being the primary camera module supplier to Nokia (enviable, that is, until Nokia's sales went into free fall in 2010). The relationship was so good that Nokia was responsible for 21 per cent of ST's annual sales of $10bn in 2007.

A shareholder presentation in July 2007 provided a snapshot of all this marketing activity. 1...Limited had identified 14 suppliers of camera modules to the 'Big Four' mobile phone vendors. Of these, 12 had asked for, and had received, samples of the SMA-based actuator. As for the manufacture of the actuator itself, Seiko Instruments had invested in a factory in Thailand. But working with Seiko had been far from plain sailing, with unresolved issues in the relationship, both technical and commercial. The mass production start date had slipped from August 2007 to February 2008.

PANASONIC AND SEIKO

Then, in October 2007, 1...Limited's shareholders were told of a fresh opportunity to enter the marketplace. One of the smaller camera module competitors, Panasonic Photo Lighting (PPL), a subsidiary of the large Japanese consumer electronics – and digital camera – company, Panasonic, was showing serious interest, targeting two main handset customers: Motorola and Panasonic's own mobile phone operation. Motorola was clearly the big prize here. Panasonic, like all the Japanese mobile companies, had become a bit player in smartphones.

Meanwhile, at 1...Limited, there had been management and financial changes. Specifically, a new CEO, Simon Calder, had been appointed and took up the job at the beginning of 2008. Tony Hooley had been CEO since the departure of Mark Shepherd. It was always intended to be a temporary measure. The role had become more commercial, and Tony was anxious to get back to research. In fact, around this time, Tony set up his own company, Hooley Research, to explore fields of interest to him. Simon, with the benefit of fresh eyes, instigated a restructuring of the business, cutting annual overheads by £1mn in 2008. As we noted in Chapter 4, CML (as 1...Limited had become) actually made a profit in 2009, helped by the reduced cost base and a large one-off DSP licence payment.

By the middle of 2008, the picture was once again mostly downbeat. But PPL, Panasonic's camera integrator subsidiary, had become a real prospect. With regard to Seiko, nearly all the technical issues had been resolved but the company was holding out for more control over the IP than CML was prepared to concede. Negotiations dragged on through the summer. Nevertheless, Seiko successfully delivered working samples to customers, particularly PPL, during this period. Simultaneously, a drive chip from a Californian semiconductor specialist called Atmel Corporation, commissioned by CML, was being tested.

On 25 September 2008, the board of Seiko Instruments finally gave the go-ahead for the installation of automated equipment for the volume manufacture of SMA actuators. This was despite the fact that the licence agreement between CML and Seiko had still not been signed. Another positive was that PPL was engaging well with both Seiko and its mobile phone customers: Motorola, Panasonic Mobile and another Japanese company, NEC.

June 2009 brought better news. After close to three years of technical development and tortuous commercial and legal negotiations, CML signed a licence agreement with Seiko. Volume manufacture in Japan was ready to go at one million actuators a month, expandable to two million a month. More significantly, PPL had obtained a design win with its sister company, Panasonic Mobile, in a handset designated the P-05B, to be launched in Japan in the first quarter of 2010.

SAES GETTERS

However, before actuator production could start, CML was faced with making another significant change. Both Seiko and Panasonic were unhappy with Dynalloy as the SMA wire supplier. The firm was unable to meet their supplier audit standards. CML was therefore obliged to move to a bigger and more capable Nitinol wire supplier, the Italian company, SAES Getters, which was – and is – one of the leaders, if not *the* leader, in this specialised field. The SAES Group (Società Apparecchi Elettrici e Scientifici – electrical and scientific components company) was founded in 1940. The impetus for the creation of the company was the discovery of a new method for the production of getters. (A getter is a substance, usually a metal, evaporated onto the walls of a vacuum space to absorb the residual gas and lower the pressure.) In 1986, SAES Getters went public on the Milan Stock Exchange. The year 2000 was marked by a new, ambitious direction intended to leverage the firm's expertise in metallurgy and materials science into related advanced materials sectors, organically and by acquisition. Shape memory wire was at the forefront of this strategy.

A key division in SAES Getters is Metallurgy. Inside Metallurgy sits a sub-division called SMA Industrial, supplying Nitinol wire for devices like actuators, valves and dispensers. SMA Industrial is small in the context of SAES Getters. Nevertheless, from CML's perspective, the SAES Getters relationship particularly appealed because the company was actively interested in promoting the usage of Nitinol wire in new applications. In pursuit of this strategy, in 2011, SAES Getters set up a 50/50 joint venture with a German automotive parts company, Alfmeier Präzision, called Actuator Solutions GmbH (ASG), to produce components based on SMA technology. As the joint press release at the time said:

> 'We both sensed the hidden need for advanced SMA actuators in various industries and the joint venture will be the first of its kind ...'

Alfmeier specialised in products that cross the boundaries between plastics, electronics and mechanical engineering and, as such, had the necessary expertise to turn Nitinol wire into actuators.

SEIKO-DRAMA

In April 2010 Seiko finally moved into mass production mode. The team at CML breathed a collective sigh of relief. The elements of the mini-supply chain that CML had so painstakingly created (wire, actuator, IC drive chip, module integration) to provide Panasonic Mobile with the advanced AF camera module it wanted all came together. The Panasonic P-05B went on sale in Japan on 24 June 2010. It featured an 8.1MP camera with 'high-speed autofocus'. Panasonic's customer was NTTDoCoMo, the dominant mobile phone network operator in Japan. The P-05B camera was well received and the phone sold well for the rest of the year, providing CML with its first royalties. By way of context, Panasonic was selling about five million phones a year at this time in Japan, small in the context of 1.6bn global mobile sales (only 30 per cent of which were smartphones). The relationship between PPL and Panasonic Mobile continued to develop well, envisaging a smaller AF SMA actuator for the P-05B replacement model, scheduled for 2011.

Unfortunately, the mobile industry background was becoming less favourable for CML. This was a time of ferment in smartphone software. The P-05B used Linux, an open-source operating system. Android was fast gaining ground on Linux, Symbian and Windows. In November 2010 Panasonic announced it would switch to Android-based smartphones in the next year and sell these models overseas in 2012. The company freely admitted it was 'late to the Android party'. Then, in July 2011, Panasonic launched the Android P-05C, with a 13.2 MP camera. It did *not* have a CML autofocus actuator inside.

What caused Panasonic Mobile to dispense with a technology that had worked well on the P-05B? In March 2011 it seemed fairly certain that the smaller AF SMA actuator that CML had developed would be incorporated into the new phone. However, as so often happens when pressure is exerted, the old technology, VCM (Voice Coil Motor), fought back and intensive efforts to reduce the cost – and the size – had borne fruit. (How this was achieved is the subject of Chapter 15.) Seiko was therefore faced with the prospect of making an actuator that was not competitive in price with that produced by the incumbent VCM firms.

On top of this, financially, 2011 was a tough year for Seiko Instruments. Profits at the Electronic Components Division fell by 79 per cent and the Seiko group as whole made a large loss. Manufacture of the SMA actuator was insignificant in the context of Seiko Instruments and represented a diversion from the mainstream, so there was limited internal commitment. Accordingly, top management decided it was no longer prepared to sanction the necessary investment for the new model SMA actuator. So, in terms of the Panasonic phone prospect, CML was left high and dry. As it turned out, Panasonic's plans to become a factor in the Android market worldwide soon collapsed. Continued deficits forced management to pull out of international sales. In 2013 the company was dropped as a supplier to NTTDoCoMo and withdrew entirely from making smartphones for the Japanese consumer market.

Fortunately for CML, in view of the way in which the VCM manufacturers had forced the price down, the mobile phone industry in 2010 was abuzz with a further technology that could be added to enhance smartphone cameras: Optical Image Stabilisation, or OIS. In itself, OIS was nothing new. Panasonic invented the original video camera with an Optical Image Stabiliser in 1988. Digital cameras routinely incorporated it to counter handshake. Like autofocus, it needed a motive force to move the lens unit. The problem, even more so than for AF, was how to squeeze a clunky VCM-type motor into a smartphone. Meanwhile, AF was starting to become a standard feature in mobile phone cameras. OIS actuators appealed to CML as they are intrinsically more complicated to make and offered the VCM makers less scope for automation and cost reduction than AF, making SMA OIS potentially more defensible.

A few observations on the events described in this chapter. First, 1...Limited, perfectly logically, focused its AF actuator marketing effort on cameras for the largest smartphone vendors. In 2007 there was no way of knowing that the mobile line-up would change so dramatically between then and 2012. By the time the actuator became available, Nokia and Motorola were fading fast. Second, the window for SMA AF closed rapidly when the VCM manufacturers raised their game. Third, Seiko turned out to be a disappointing partner. In any event, CML became enamoured with the thrust into the uncharted territory of OIS. What happened next is the subject of Chapter 11, after we learn how the key bits of a camera module fit together.

10.

Camera modules: integrators, sensors and lenses

CAMERA MODULE INTEGRATORS: OVERVIEW

The sorry Seiko–Panasonic saga described in the last chapter highlights the importance of being able to negotiate the tricky set of relationships between component makers and camera module integrators. The list of module manufacturers may have changed since 2010, but the issue today is exactly the same: how to persuade a module integrator to adopt *your* actuator for the next model of handset. Something that is inevitably more difficult if you are promoting a new and unfamiliar technology.

We talk colloquially about a smartphone's camera, as if there was only one rear camera. This is a carryover from digital cameras, which have a single image sensor and multiple, interchangeable lenses. The reality is that, since 2017 when multi-camera phones first emerged, 'the camera' consists of several camera modules, each with its own sensor. Camera modules go into everything, from laptop computers to security systems. The smartphone is only one of a wide range of applications. It is, however, the most demanding in terms of performance. Generally, phones produced nowadays contain three rear-facing cameras and one front-facing (selfie) camera. Each one is a module. In round terms, therefore, with around 1.2bn phones sold annually (many with fewer than three rear cameras), camera module demand is of the order of 4.5bn units a year.

What are the components that the module integrator has to assemble to create a camera module for insertion into a handset?

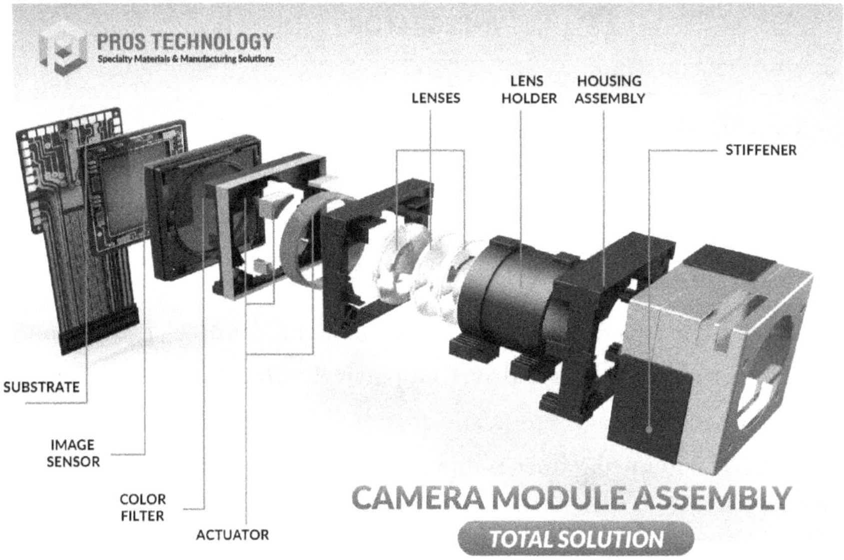

Source: Prostech

The above diagram shows the parts of a camera module: image sensors, lenses, actuators, filters, etc. Adhesive is used to bond these parts together. The two key elements – the image sensor and the lens unit – are located at opposite ends of the module. These elements account for the bulk of the cost. The third most expensive item is the actuator. How are all these parts brought together to form a module? The following diagram traces the flows.

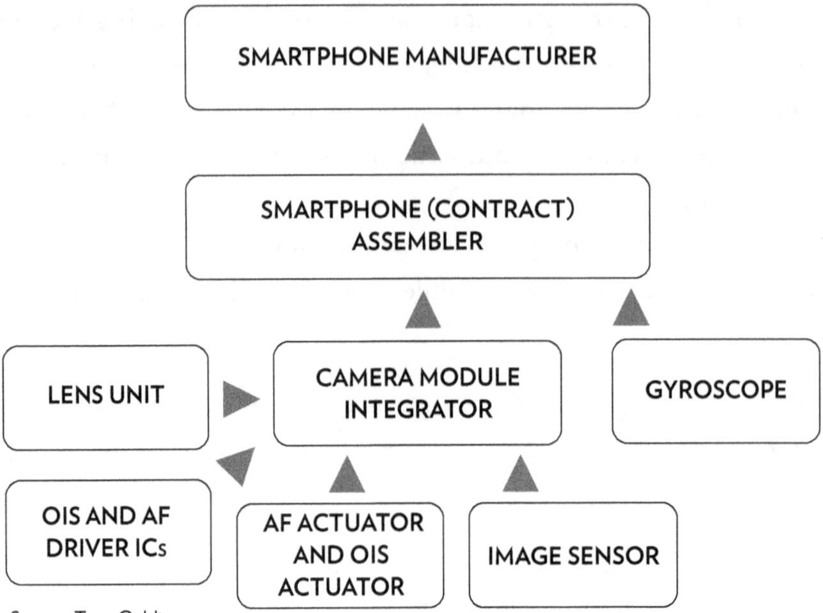

Source: Tony Golding

Separate OIS and AF driver ICs are needed to control the actuator mechanism. Typically, the driver chips are located inside the camera, quite close to the image sensor, and are not visible from the outside of the module. The following image shows just how small driver chips can be.

Source: Cambridge Mechatronics Ltd (CML)

All of the terms in the schematic above, showing the delivery flows, will be familiar by now, except gyroscope. That is because the gyroscope is an integral part of the OIS system. (The role of the gyroscope is covered in the next chapter.)

A good deal of vertical integration – making your own module components – occurs in this field, especially among the larger contenders. The table below characterises the camera module industry in these terms, ranking the top five by their estimated position in the smartphone marketplace.

MODULE INTEGRATOR	ORIGIN	Market rank	Module assembly	Makes lens units	Makes image sensors	Makes actuators
LG Innotek	South Korea	1	✓	✓		✓
Sunny Optical	China	2	✓	✓		✓
Samsung (SEMCO)	South Korea	3	✓	✓	✓	✓
Sharp (Foxconn)	Japan/Taiwan	4	✓	✓		
OFilm	China	5	✓	✓		

Samsung Electro-Mechanics (SEMCO) is firmly in the 'do as much as possible internally' camp, as is LG Innotek. The firms above that are vertically integrated do not necessarily buy from sister subsidiaries. In the module business – and in the smartphone business generally – there is much cross purchasing: the buying of components from outside your internal supply chain. Thus co-operation and competition co-exist – happily or not so happily – between participants. (Appendix 4 contains several examples). Smartphone makers are increasingly encouraging camera module integrators to vertically integrate, in the belief that a single, co-ordinated entity offers a better and cheaper product. The complex semiconductor technology involved in image sensors means that this technology is beyond the capability of module assembly companies but VCM assembly is not. Consequently, LG Innotek, Sunny, SEMCO and others manufacture VCMs.

Sunny Optical, Chinese owned and quoted on the Hong Kong stock exchange, is both a lens manufacturer and a module integrator and runs these businesses in parallel. Sunny uses some of its lens units internally and sells the rest to third-party integrators. A rising star in the camera module industry is Q Technology, a Chinese company quoted in Hong Kong with ambitions to become a vertically integrated supplier. Beyond these, the industry mostly consists of smaller Chinese, Korean and Taiwanese contenders that do not make any of their own components.

Apple is, of course, a large purchaser of camera modules and has, traditionally, split its business between LG Innotek, Sharp (controlled by Foxconn) and the leading Chinese integrator, OFilm. However, in late 2020, Apple dropped OFilm due to its alleged use of forced labour, and the two remaining suppliers were asked to fill the gap. LG Innotek is the market leader in camera modules for smartphones, with a share of approximately 20 per cent. The other side of the coin is its extreme dependence on Apple, something that troubles shareholders (although part of the LG Group *chaebol*, LG Innotek is listed on the Korean Stock Exchange).

With the advent of multi-camera phones after 2016 – a pivotal moment in smartphone photography – the possible number of rear camera modules on a phone proliferated. The different types of smartphone camera are listed here. They are fully explained in Chapter 20.

- » Primary/wide-angle lens
- » Ultra-wide-angle lens
- » Zoom (or telephoto) lens
- » Periscope (or folded) lens
- » Monochrome lens
- » Macro lens
- » Time-of-Flight (ToF) camera

Designers of premium phones combine four or maybe five of these types of camera modules to provide the user with a coherent photographic experience, consistent with space and cost. A popular combination is three rear-facing camera modules: wide-angle (the primary camera), telephoto and ultra-wide-angle. Quad camera phones usually add either a periscope camera, a monochrome camera or a Time-of-Flight camera. The image below shows a four-camera module line-up.

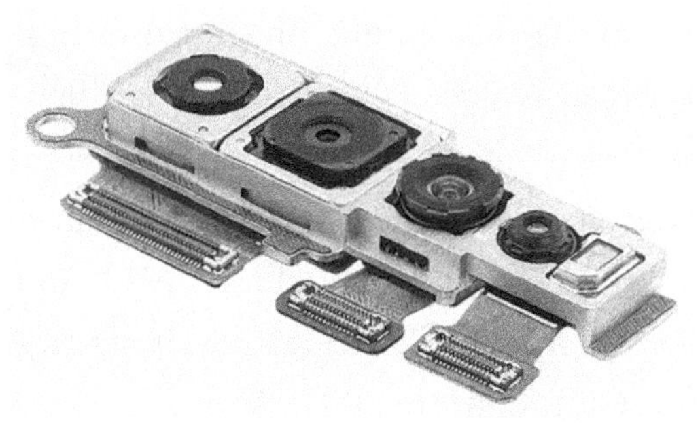

Source: Samsung Electro-Mechanics Company (SEMCO)

The years after 2016 witnessed the flowering of smartphone photography as manufacturers explored the boundaries of a multi-camera world. What happened in those years is the subject of Chapter 20.

IMAGE SENSORS

The most expensive component in a camera module is the image sensor. In revenue terms, the technically sophisticated smartphone CMOS (complementary metal-oxide semiconductor) image sensor business dominates the image sensor market. Laptop cameras, for example, are crude by comparison: low resolution, less than 5MP and often less than 2MP. The manufacture of image sensors is a niche sector of the semiconductor industry. According to the Semiconductor Industry Association, image sensors only account for

3 per cent of total industry turnover. It is, however, a big niche, with sales in the region of $20bn annually.

Image sensors come in different sizes. For historic reasons, the physical size of a sensor is expressed in imperial rather than metric units. It is expressed as a fraction, in inches: as in ⅓" and 1". Fortunately, these fractions have metric equivalents, which are instinctively easier to absorb. A ⅓-inch sensor has an approximate 6mm diagonal, while a 1-inch is close to a 16mm diagonal.

Sony got into the smartphone image sensor business early and now possesses a broad and deep technological base that others struggle to replicate. In the early days, CMOS image sensors were inferior to the CCD (charge-coupled device) image sensors in digital cameras. However, Sony could see the potential for CMOS in phone cameras and shifted resources from CCD to CMOS, launching its first CMOS image sensor for smartphones in 2007. With due acknowledgement to Jane Austen, it is a truth universally acknowledged that, in the field of smartphone image sensors, Sony has, for more than a decade, enjoyed technological leadership, although Samsung is a competitive No. 2.

In terms of revenue, Sony accounts for more than half the sensors that go into camera modules for smartphones. The company's 'anchor' customer is Apple. It has long supplied 100 per cent of Apple's needs, though second-placed Samsung threatens to erode its monopoly. OmniVision, which, despite its Western-sounding name, is Chinese-owned, lies third. An additional Chinese contender, SmartSens, is carving out a name for itself in smartphones. Further down the pecking order is Chinese image sensor maker, GalaxyCore, which mostly manufactures inexpensive sensors with resolutions below 8MP, for entry-level smartphones and other applications. SmartSens and GalaxyCore will both figure in later chapters. Smaller image sensor makers tend to be fabless but there is a trend towards building your own fab to ensure control over wafer supply.

Where do image sensors fit within Sony? It is, after all, a multifaceted conglomerate, encompassing games, music, films and electronics. The dynamics of the Imaging and Sensing Solutions (I&SS) division – accounting for 13 per cent of group sales – are different from the rest of the business. Seventy-five per cent of the sensors the division makes go into mobile phones. Sony fabricates most of the image sensors it sells in its own semiconductor facilities. Image sensors do not require state-of-the-art process nodes – 22nm and above will suffice.

Sony originally viewed image sensors as a rising star in its corporate firmament but, since 2020, the division has lost its lustre. Despite the continued growth in demand for premium phones, IS&S has been held back by the stagnant handset market. Operating margins have more than halved due to cost pressures and excess production capacity. In 2025, it emerged that Samsung had breached Sony's iPhone monopoly. Apple has plans to introduce Samsung sensors in future models. There are persistent rumours that Sony will spin off its image sensor activities into a listed business, retaining a minority stake. IS&S is heavy on R&D and capital expenditure, and it makes sense to offload some of the responsibility for funding these onto external shareholders.

From the early days of camera phones, vendors encouraged buyers to focus on the number of megapixels offered by a camera's CMOS image sensor as a single metric for camera quality. In truth, this is far from the case. There are many other factors. Apple, to its credit, refused to be drawn into megapixel mania, retaining its highly regarded 12MP sensor cameras for several years while its competitors were noisily shooting for 100MP or more. With the introduction of multi-camera phones from 2017, the megapixel race cooled for a time. It was then revived, spearheaded by Samsung, Xiaomi and Motorola. The reason for this emphasis on megapixels is surely marketing. A big number plays better with the buying public than a small one – more must mean better! In practice, it is not as simple as that. Chapter 20 explains why.

Digital cameras have lenses that can physically zoom in or out (optical zoom) to vary the focal length and capture different kinds of shots. Applying this technology to smartphones would result in a large protrusion on the back – which would be unacceptable to the consumer. To overcome this, and keep their devices thin, smartphone designers have instead gone for a combination of different fixed focal length lens units.

Progress in image sensors continues but smartphone designers are up against the laws of physics – the limits imposed by the dimensions of a handset. In hardware terms, there is no way in which a smartphone can totally match a traditional digital camera. Then, starting in 2017, several trends came together. First, the emergence of multi-camera phones. Second, SoCs with greater computing and memory capability. And, third, sophisticated software incorporating more complex algorithms. Indeed, in the last few years, advances in software have been behind the impressive improvement in the performance of smartphone cameras. This expanded ability to enhance the quality of the images taken by a smartphone camera is called computational photography.

Computational photography is a critical and generally unsung element in mobile phone cameras. Software algorithms work in tandem with image processing. Basically, camera developers found a way of using a smartphone's strengths – fast electronic shutters, powerful processors and software – to overcome its weaknesses. This graphic from a specialist photography website sums up crudely but effectively how the electronics and the software embedded in a phone compensate for the sheer physical capability of a traditional DSLR (Digital Single-Lens Reflex) camera.

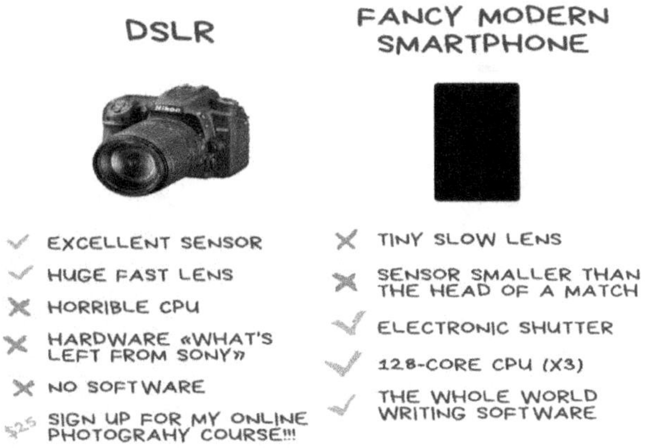

Source: 'Computational Photography', Vasily Zubarev, DP Review, 3 June 2020

LENS UNITS

A key component in a camera module is the lens unit, through which the light enters to hit the image sensor. A smartphone lens unit will contain several separate lens elements (typically between four and seven), each of which is a lens in itself. Why so many lens elements? Lens elements have curved surfaces that reasonably accurately project images onto curved surfaces, although each will have its own distortions. Several stacked lenses are needed to remove the geometric and chromatic distortions when projecting onto a flat surface. The tiny lenses in today's smartphones are stacked several millimetres high, as in the graphic on the next page, which is one reason for the rear camera bump.

The extent to which the lens opens to let in light is known as the aperture. Apertures are expressed as a number known as f-number or f-stop, with the letter 'f' appearing before the number, like f/8. Confusingly, small numbers represent large apertures, and large numbers represent small apertures. The other key factor is the time the lens is kept open – the shutter speed. Clearly, a longer exposure time is better, as it lets in more light.

Source: Sony, in 'Smartphone imaging technology and its implications', published by De Gruyter Brill, 10 August 2021

Contrary to popular perception, the vast majority of the optical lenses in mobile phone cameras are made of plastic, not glass. Plastic lenses are not as good as glass lenses but are considerably cheaper. As plastic lenses are easily injection moulded, they are reproducible in the millions on automated lines. Glass is also a great deal heavier than plastic, needing a larger actuator force to move it. To improve picture quality, a few premium smartphone models have introduced hybrid lenses – replacing one or even two of the multiple plastic lens elements with a glass lens. This approach has gained little traction so far.

As a consequence of the wholesale camera migration from Japan to elsewhere in Asia, the manufacture of lenses for smartphones is dominated by a dozen companies based in Taiwan, China and South Korea. The undisputed leader is Largan Precision which is quoted in Taiwan. It is widely considered to be the premier lens company, in terms of both value of sales and technological capability. Largan accounts for the lion's share of Apple's business and controls maybe 35 per cent of the smartphone lens market by value. Founded as a glass lens manufacturer, Largan – with commendable foresight at the time when smartphones became important – took the bold step of pouring all its resources into the efficient manufacture of plastic lenses. The second ranking lens producer is Sunny Optical. Sunny essentially serves the Chinese smartphone market though it has recently qualified to supply Apple. Third is Genius Electronic Optical, another Taiwanese producer, Apple's second source lens supplier.

11.

Smartphone makers say yes to OIS

OIS, EIS AND GYROSCOPES

By September 2011 CML had regrouped after the Seiko–Panasonic disappointment. The positive from the experience was that the company had demonstrated that SMA autofocus actuators could be mass produced. The bad news was that, despite the rapid market expansion, autofocus was not an attractive market to be in. SMA had been vanquished by VCM on cost grounds. Although efforts were made to develop a more competitive SMA AF product, essentially CML set its sights firmly on Optical Image Stabilisation (OIS), which, as we saw at the end of Chapter 9, had emerged as the Next Big Thing in mobile photography.

What exactly is OIS? Digital cameras have offered stabilisation techniques as a feature since the mid-1990s, often giving it a distinctive name to associate it with their own brand. For example, Sony designated it Optical SteadyShot while Minolta simply called it Anti-Shake. Just as with autofocus, mobile phone companies were keen to transfer the OIS function to smartphone cameras. As the number of megapixels grew, and resolution improved, there was an increased need for stabilisation. In addition, camera module integrators were already envisaging a zoom capability, bringing with it a greatly increased handshake risk. OIS was an important step on the road to persuading keen photographers to use their phone camera instead of their digital camera. There is a well-worn – though no less true – aphorism in photographic circles that 'the best camera you own is the one you have with you'. Why carry a bulky digital camera if the one in your pocket is just as good?

In colloquial terms, OIS is 'anti-shake'. It attempts to counter the unsteadiness of the human hand and the resultant blurry, ill-defined shots. Other image-stabilisation techniques exist, specifically EIS (Electronic Image Stabilisation) which relies on the computational power of the phone. By contrast, OIS uses hardware built into the camera module to counteract hand movements while taking a picture. OIS moves the lens – or indeed the image sensor or module itself – incredibly rapidly to compensate for shake. As with autofocus, an actuator is needed to accomplish this. Although they appear to be in conflict, OIS and EIS are actually complementary and today's high-end phones normally offer both, operating in tandem. It is important to note that not every camera on a multi-camera phone will have OIS and/or EIS.

EIS reduces image shake effects by manipulating electronically the successive frames on the image sensor. This is achieved by software. As such, EIS is a constituent part of computational photography. EIS relies on a motion sensor to track the source of the jitter, such as a shaky hand or vehicle motion. Whether taking photos or shooting video, the camera continuously captures frames. EIS crops and shifts these frames to counteract the detected movement. For example, if the camera shakes to the left (exposing more of the image to the left) then the left part of the frame is cropped to compensate (and vice versa). The downside is that, as the cropped frames occupy only around 90 per cent of the sensor, when the image is electronically magnified to fill the remaining 10 per cent of the frame, there is a small loss of image quality and resolution.

In contrast, OIS physically moves a part of the camera, normally the lens, in response to any shaking detected by the phone. This mechanical movement, however, takes time, which is a problem when shooting high frame rate video. Consequently, there is a consensus view that EIS is more effective for stabilising video. Conversely, OIS is considered to be a superior stabilisation method for still photography. This is true especially in low lighting conditions as, to achieve a good picture, the aperture has to stay open – in photographic terms – for a long period. OIS captures sharp pictures at shutter speeds three, four or five times slower than otherwise possible.

In VCM-based OIS, two components work in tandem with additional electronics to feed the actuator the necessary instructions:

» A gyroscope is used to sense vibrations or movement of the device while taking the picture. The gyroscope measures angular movements of the camera which are caused when hands shake while clicking. As well as their role in OIS, gyroscopes are used in smartphones to enable commands via deliberate shaking and for motion gaming, such as when using the phone as a steering wheel.
» An actuator receives data from the gyroscope. VCM actuators normally require a position sensor for accurate control. SMA actuators do not need a separate position sensor as the SMA wire acts as both actuator and sensor.

Most OIS motion control systems use a feedback loop. This involves taking measurements of the phone's position with a sensor and making a calculated adjustment to any position error. This error signal is sent to a microcontroller, which sets the desired new position. The microcontroller is usually included as a function on the driver chip. The resulting data stream is then fed to the driver IC, which instructs the actuator to move the lens to compensate for the hand jitter. OIS is a considerably more complicated set-up than autofocus.

Gyroscopes (or gyros, for short) are a key element in OIS. However, they are just one of the motion sensors – the generic name – to be found in a smartphone. Another common motion sensor is an accelerometer. Both are inertial sensors – transducers that convert inertial force (the physical force that an object experiences when being accelerated linearly or rotationally) into electrical signals. Accelerometers measure, for example, steps, whereas the gyro senses small movements (such as tilting the phone) that do not actually cause it to move through space.

Gyroscopes were first used in digital cameras in the mid-2000s. They were first applied in smartphones in 2009. For this to happen, gyros had to be made smaller and more robust. Fortunately, a novel production technique called MEMS, based on wafer fab, semiconductor-type processing, had been developed that was able to do precisely this. MEMS gyros (and other components) have been used in handsets since then. MEMS is an abbreviation of micro electro-mechanical system. MEMS-based sensors, like gyros and accelerometers, are widely used in modern electronics. Most microphones in consumer electronics products have been MEMS-based since 2010. MEMS is an ingenious production technique that mimics many aspects of the semiconductor fabrication process, specifically silicon-based batch production. While the device's electronics are fabricated using IC technology, MEMS permits the construction of tiny mechanical components alongside, fabricated by the sophisticated manipulation of silicon and other substrates using micromachining processes.

As the diagram below shows, in a camera module, the OIS actuator sits behind the lens and the autofocus actuator and before the image sensor. In terms of function, the key element in either an SMA AF or OIS actuator is the Nitinol wire that lengthens or shortens in response to a small change of current generated by the driver chip.

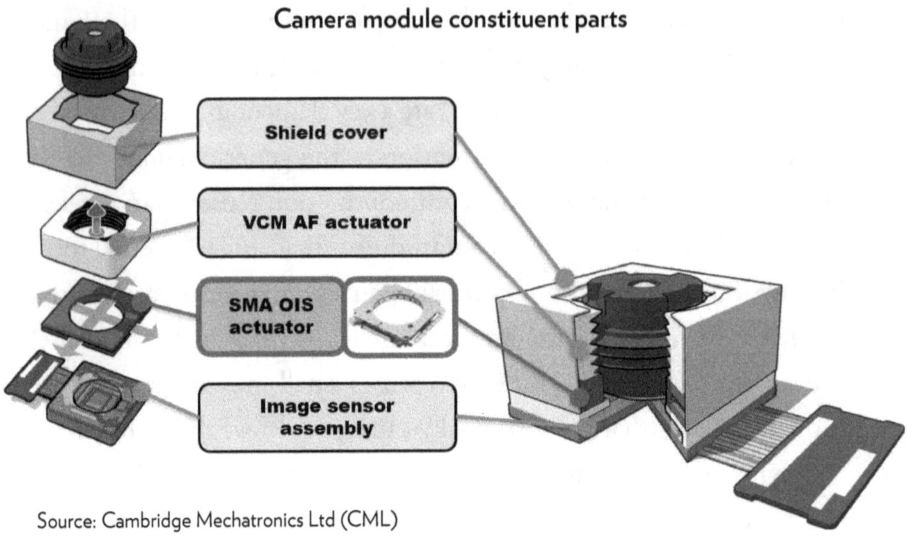

Source: Cambridge Mechatronics Ltd (CML)

An SMA autofocus actuator operates in just one dimension and therefore requires a single SMA wire (or sometimes two) while an OIS actuator needs four as it has to operate in two dimensions. In an OIS unit, each of the four SMA wires is individually addressable by the driver chip. By sending a different current down each wire, the actuator can be moved in any direction. Within CML, the shorthand for this is '4-wire'. SMA OIS was deliberately designed to work in conjunction with an existing VCM autofocus actuator in a camera module. Module makers were therefore free to retain their current VCM AF supplier and add this new type of OIS.

Looking ahead, CML could see a path to combining AF and OIS in a single unit using eight SMA wires ('8-wire'), which would lead to a reduction in both size and cost. In principle, it was clear that an SMA OIS actuator had fewer parts and should be cheaper to assemble than the equivalent VCM actuator. VCM autofocus had been largely automated but the substitution of equipment for labour was more difficult in respect of OIS actuators. For one thing, there were many more parts to assemble. It certainly looked more defensible.

GETTING INTO POSITION FOR THE OIS GOLD RUSH

During 2012, in mobile phones, the race to embrace OIS had some of the characteristics of a gold rush. Everyone agreed it was going to happen. The consensus was that it would start in 2013 with real volume in 2014. For CML, SMA offered the advantage of a level playing field. As the OIS function did not currently exist in phones, there was no embedded actuator technology to overcome. It was therefore an opportunity for CML – or anyone else – to insert something new. CML had proved SMA actuators could be mass produced and had commissioned two ASIC control chip drivers from fabless semiconductor firms to work in conjunction with them.

Nevertheless, doubts remained in the minds of smartphone makers and camera module makers about the company's ability to ramp up production via its licensees. To this end, CML talked to ASG, the SAES-Getters–Alfmeier joint venture, about a stronger relationship than that of licensee, more of a manufacturing partnership. The problem with the Seiko relationship was that CML's role was not much more than that of an introduction agency. After the initial connection, the two parties – licensee and module maker – were just left to get on with it. A manufacturing agreement would give CML greater control over the technology. Better still, it meant that the company could effectively offer the camera module maker a component rather than just a relationship.

In an effort to link all the elements in the camera module chain, CML had extensive discussions with Foxconn, then the second largest global camera module integrator, with an established position in the Chinese market. OIS additionally opened up a major opportunity to connect with the leading smartphone vendors. In 2012 Samsung and Nokia controlled over 40 per cent of the global market between them. But – and it was a big but – Nokia's star was fading fast while Apple was in the ascendant. Samsung, with an annual output of close to 400mn handsets, and Apple, shooting for 200mn, were clearly prime targets. Discussions on SMA were initiated with Samsung. They proceeded slowly as Samsung showed a marked reluctance to be a first mover in OIS. CML talked to both Samsung Mobile and Samsung's camera module integrator affiliate, SEMCO. In the event, these discussions were discontinued. Both parts of Samsung decided not to proceed with SMA OIS.

Then, on 30 March 2012, CML disclosed to shareholders that it was in commercial discussions on exclusivity with a 'Tier 1 mobile phone handset manufacturer'. The identity of the company was not revealed. All went quiet until, suddenly, in December 2012, shareholders were told that discussions with the Tier 1 phone manufacturer had been terminated. The indications are that the company in question was Apple. Richard Topliss, CML's lead inventor of SMA actuators, had been recruited by Apple as a Senior Camera Technology Specialist in 2011 and started work in California in January 2012. His departure, though undoubtedly a blow, would have given CML an entrée. That the discussions were 'robust' – to use the standard euphemism – becomes apparent from this quote from Dr Martin Knight, who had been elected chair of 1...Limited's board in 2004, in a shareholder letter dated 30 September 2012.

> 'There are many reasons why it is taking time to persuade the market to take our technology on terms attractive to us. Shareholders will be aware of the belligerence of the industry giants we are negotiating with and indicative offers for our technology have been unattractive.'

Presumably, in deciding whether to hitch its star to one of the smartphone industry's major participants, the following considerations would have figured prominently. Any insistence on exclusivity – preventing CML from selling to any other smartphone manufacturer – would have acted to curb the company's wider market potential and leave it dependent on a single customer and its whims. Additionally, it was highly probable that there would soon be pressure for a second source to which CML would have to transfer its hard-won expertise. The company's control of its proprietary SMA technology could easily slip away.

Further to this, CML had ambitions for SMA beyond mobile devices that could be placed in jeopardy. Having spent much time and engineering resource mastering and patenting SMA, CML viewed its technology as a *platform*, a base from which it could move into additional applications. This was Tony Hooley's philosophy from the start, and it had become imbued into CML. Doing a deal with just one company would not per se interfere with the pursuit of non-smartphone opportunities but there was a danger that constant demands would consume engineering resource and reduce the company's ability to diversify.

As 2013 dawned, CML found itself without the high-volume launch customer it had hoped to secure. To have convinced either Samsung or Apple to go with SMA for OIS – on terms acceptable to CML – would have been a tremendous coup and set the company on a path to profitability. Even a single model from either of these two industry giants would have produced initial orders for several million actuators.

Other possibilities arose in 2013. For a brief moment, the Canadian smartphone producer, Research in Motion (which in that year renamed itself BlackBerry, the name of its product) was the front runner to adopt SMA OIS. SMA was firmly in the running for the OIS actuator. But it was too late. BlackBerry survived but not as a smartphone manufacturer. For CML the timing was singularly unfortunate.

ENTER HUTCHINSON

Negotiations with possible manufacturing partners continued through 2012 and into 2013. ASG was quickly ruled out of the running due to lack of experience in volume production. Management spent time evaluating the kind of company that would be a suitable manufacturing partner. In late 2012, a chance meeting between one of CML's engineers with his equivalent in a small American firm, Hutchinson Technology Inc., threw up an opportunity. HTI was located in the middle of rich farmland just west of Minneapolis in the little town of Hutchinson, Minnesota (population 15,000). Although quoted on Nasdaq, it was lowly valued with a market capitalisation of a mere $80mn, loose change in American stock market terms.

Hutchinson manufactured suspension assemblies. A suspension assembly is a component in a Hard Disk Drive (usually abbreviated to HDD). In 2013 the majority of personal computers were sold with HDDs inside. Suspension assemblies are a high-volume, precision product – a thin metal spring that holds the recording head in position a few millionths of an inch above the surface of a rapidly spinning disk. HTI manufactured these close-tolerance components in quantity, using chemical, mechanical and electronic technologies. It soon became apparent to CML that the firm possessed the necessary skills and equipment to manufacture SMA actuators. The company was loss-making and debt was mounting. With solid-state drives (SSDs) starting to present strong and growing competition to HDDs, what Hutchinson urgently needed was an additional product to fill its underutilised production lines. CML offered just that.

In April 2013, after six months of familiarisation, the two companies entered into a collaboration agreement. In practice, it was more of a joint venture. Hutchinson became CML's principal manufacturing partner. The two companies worked together to refine the actuator design and improve manufacturability. The challenge for CML and HTI was to construct a device of acceptable quality and then – crucially – to ramp up production to the point where the company could reliably deliver millions a month.

PART 3

CHIMING WITH CHINA

12.

Pivot to China

RISING OPPORTUNITIES IN THE EAST

Unable to secure a customer for its OIS (Optical Image Stabilisation) actuator, from another perspective CML was lucky in its timing. In 2013 Chinese smartphone producers, along with Apple, began to make their mark in world markets, filling the void left by Nokia and BlackBerry. Foxconn lost a contract to make camera modules for the iPhone and found itself with spare capacity. It set out to attract new smartphone customers, thus offering CML the chance to obtain design wins for its actuator.

As a Taiwanese company, with numerous plants in China, Foxconn had existing relationships with various Chinese mobile customers, notably Huawei, Coolpad and InFocus. Huawei was rising fast at this point. It had transformed its consumer division from a manufacturer of handsets for telecommunications providers into a strongly branded operation with its sights on the upper end of the smartphone market. Coolpad had risen from nowhere to become the third largest selling brand in China.

InFocus was a curious participant in the smartphone market. Its handsets were made by FIH Mobile, a 66 per cent owned subsidiary of Hon Hai, Foxconn's parent company. FIH also assembled phones for Nokia and Motorola and, as these brands lost market share, it lost money. The partnership with Foxconn was a strange one because InFocus was – and still is – a private American company, with its headquarters in the small city of Tigard in Oregon, specialising in the sale of office and home projectors. In 2009 it was acquired by John Hui, a Chinese American entrepreneur, who grew up in Hong Kong and then made his fortune in the US tech industry. Foxconn did not own shares in InFocus: it was a collaboration. In practice, there is little doubt that Foxconn was in the driving seat. Quite why it teamed up with a company with no experience of the mobile market is a mystery.

As for supply of actuators, CML spent 2013 helping Hutchinson Technology Inc. (HTI) get into production. This entailed integrating the actuator into the Foxconn camera module, supplying prototype samples to prospective customers and ensuring that the final product was able to pass all the necessary performance and reliability tests. By January 2014, Huawei, Coolpad and InFocus were evaluating SMA OIS for incorporation into handsets due to be launched later that year. Foxconn was not, of course, granted exclusivity, so CML maintained its marketing in parallel to competitor module integrators. It started discussions with Primax of Taiwan. A major attraction was that Primax was shipping camera modules to Xiaomi. Sunny Optical also expressed interest in a design using CML's actuator.

On the supply front – as smartphone customers preferred not to rely on a single source – CML welcomed an approach from New Shicoh, the largest Chinese VCM actuator maker, for an SMA licence. To get New Shicoh into the market, CML began to supply the firm with OIS units. As evidence of its intent, New Shicoh designed a slim AF actuator to be compatible with SMA OIS. A little later CML was approached by TDK looking for a similar arrangement to supply its AF actuators in conjunction with CML's OIS units. Each of these micro-chains included the module integrator's favoured driver IC supplier: the American semiconductor company, Onsemi, for

Foxconn; and Renesas of Japan for Primax and Sunny. The following table summarises what these emerging supply chains looked like from CML's point of view.

OIS ACTUATOR MAKER	AF ACTUATOR MAKER	CAMERA MODULE MAKER	DRIVER IC MAKER	POTENTIAL SMARTPHONE CUSTOMERS
CML/HTI	New Shicoh	Foxconn	Onsemi	Huawei, InFocus, Coolpad
CML/HTI	TDK	Primax/ Sunny	Renesas	Xiaomi and others

Source: Tony Golding

All was set fair for a modest uplift in actuator royalties later in 2014. The projected volumes were not enthralling – hundreds of thousands per model rather than millions – but it was early days in the roll-out of OIS. Then, for 2015, assuming the momentum was sustained from new models and additional customers, internal projections envisaged 44mn actuators being shipped, generating £10mn in sales. Even Samsung, having dragged its feet on OIS, revealed in late 2013 that it would join the race with a VCM solution. Nevertheless, productive conversations continued with Samsung Mobile on SMA.

We touched earlier on the fact that OIS had an additional dimension as compared with AF: the need to connect with a gyroscope. The development of MEMS gyroscopes allowed OIS to happen in smartphones. Compatibility was essential, giving rise to a need to establish working relationships with all the companies involved, including CML. A director at InvenSense, one of the leaders in the gyro business, described the position succinctly:

> 'It is a really complex and interdependent ecosystem – OIS gyroscope, image sensor, lens and actuator with a companion control

> and driver chip, all in the most compact housing achievable.' ('How technology is bringing sharper images to users of smartphone cameras and videos', *New Electronics*, 27 May 2014)

The two pioneers of MEMS gyroscope technology for phones were STMicroelectronics and InvenSense, a newcomer based in Silicon Valley. InvenSense, a New York Stock Exchange-quoted company, delivered its first sensor in 2006. As a fabless semiconductor company, it used either TSMC or GlobalFoundries to fabricate its devices. In 2011 the company introduced a gyroscope for OIS in smartphones and, in the next two years, made inroads into Android-based handsets, with Samsung as the lead customer. Sales and profits rose accordingly, and the stock price followed suit, hitting \$25 in 2014, well up on the Initial Public Offering (IPO) price of \$7.50.

By the middle of 2014 CML could congratulate itself on having achieved much. It had three end customers in sight: a design win from InFocus and prospects for business from Huawei and Coolpad. A presentation to shareholders in June 2014 projected the familiar 'hockey stick' uplift in sales in 2015 and 2016, based on some bold assumptions. Not only did CML envisage securing a good share of the Chinese OIS actuator market, but the 2016 projection included a significant penetration of Samsung's forthcoming Galaxy S7 handset, with a projected volume of 50mn actuators a year.

Assumptions were made on production and cost. In mid-2014 the market price for an OIS actuator was \$4 (compared with \$1.40 for an autofocus actuator). As with AF, it was expected that this figure would fall as volumes increased. HTI's production cost was above this but it was confident that automation, increased volumes and improved yield would bring this down to well below \$4 by the year end. By the first quarter of 2015 it was predicted to be below \$2 and then fall in time to \$1. Inevitably, this would require further capital investment from both shareholders, CML and HTI. If all this came to pass – and the risks were not glossed over – CML would soon be making handsome profits for its investors.

SMA REACHES THE MARKET

SMA OIS finally made its debut in the InFocus M530 in January 2015. Powered by a MediaTek SoC, the M530 featured a 13-megapixel camera. The arrival of the M530 was the cause of much rejoicing within CML (and among its shareholders!). Reassuringly, sales of 100,000 handsets in the ensuing months revealed no reliability or performance issues. SMA OIS worked in a real-world environment just as expected. InFocus (effectively Foxconn) told CML that they planned to roll out more models with SMA OIS.

Huawei remained keen to press ahead with two SMA-enabled handsets for launch in the second quarter of 2015 and a delegation from the company visited CML in Cambridge. However, CML was told there would be no design win until it was absolutely confirmed that the OIS in the M530 had no reliability issues. Coolpad never materialised as an opportunity. The company's reputation was tarnished by the revelation that the software on many of its phones included a 'backdoor', enabling the installation of unwanted applications and push-notification advertisements.

How did the roll-out of OIS-enabled phones unfold in practice? It was slower than expected. A survey of the leading brands in May 2015 found that ten flagship phones were offering it. This list included two from Samsung and one, the iPhone 6 Plus, from Apple. Both had stuck with VCM for the actuator mechanism. Samsung Mobile used an actuator from SEMCO while Apple retained its two existing (Japanese) AF VCM suppliers, Mitsumi and Alps, to manufacture its OIS actuators.

As 2014 wore on the news deteriorated. Foxconn told CML there would be no follow-on design wins. A major reason given was that the actuator as it stood was too large in terms of thickness, causing the camera to stick out. A decision was made by CML and HTI to redesign the device. The new version was designated Gemini. The second-generation version would be 70 per cent thinner, require fewer components and be easier to make. But it would take time – up to a year.

The redesign was a setback but there was promise on the Samsung front. After a shaky start, the on-going conversations initiated by Samsung in early 2014, had evolved into something altogether more substantial. Meetings at various levels were held in South Korea and the UK, including with senior executives at Samsung Mobile. CML was asked for, and delivered, samples for testing during 2014 to assess their suitability for future models. These conversations were wide-ranging and envisaged the possibility of a broad collaboration between Samsung and CML. Specific projects were discussed. Samsung was considering OIS for front, 'selfie', cameras and the concept of a – smaller and cheaper – combined AF+OIS SMA actuator greatly appealed. What happened next to the Samsung relationship is covered in Chapter 17.

CAPITALISM WITH CHINESE CHARACTERISTICS

The rapid ascent of Chinese smartphone vendors as serious competitors to Apple and Samsung after 2015 could not have occurred without access to sources of funding. Where did the money come from? The answer has its roots in the 1970s when Deng Xiaoping, the then leader of the People's Republic of China, initiated reforms that led in the direction of a market economy. The following quote from an academic journal provides a useful summary of quite how profound the transition from the *dirigiste* policies pursued by the Chinese Communist Party (CCP) to free market capitalism has been.

> 'Since 2000, there has been a drastic shift in the initial public offering market across the world. While IPO volume has dropped considerably in the U.S. and other developed markets, China has experienced high IPO volume and a tremendous growth in listings. China had no stock exchanges between 1949 and 1990; yet as of 2020, it has the second largest stock market in the world, after the U.S. … During 2011–2018, 1,546 companies went public in China. In comparison, 971 operating companies went public in the U.S. …' ('Initial Public Offerings Chinese Style', *Journal of Financial and Quantitative Analysis*, July 2022)

Allowing enterprising individuals to create companies and become rich, while benefiting the economy at the same time, is a defining characteristic of capitalism. To stimulate entrepreneurship and innovation, the Chinese authorities consciously tried to replicate the structure found in Silicon Valley and similar geographical clusters, with the express intention of building a home-grown startup culture.

Venture capital (VC) is a branch of private equity, the umbrella term for investment in non-listed companies, be they startups or existing businesses. In the USA and elsewhere in the West, institutional investors buy stakes in funds run by venture capital firms which, in turn, invest in new and up-and-coming companies offering innovation and the promise of rising profit. The funds run by VC firms commonly focus on particular industry sectors, such as e-commerce, semiconductors or biotech. Often, nascent companies require several refinancings via rounds of investment. The objective is an 'exit', normally via a sale to another company or an IPO. Most of these aspiring companies fail, of course, but the few that do succeed have a marked positive impact on the economy and society.

Starting in the late 1990s, the Chinese government set up government-guided investment funds (GGIFs): public-private investment vehicles with both financial and industrial policy objectives. China had adopted the Western venture capital template, but the authorities determined it was to be a model with 'Chinese characteristics'. GGIFs gained little traction in the early years as the VC market boomed, with international institutional investors, principally American, playing a leading role. In 2010, in a bid to construct a domestically controlled venture capital system, the Chinese authorities re-established GGIFs on a more formal basis, using a beefed-up fund structure.

These new and enhanced guidance funds soon expanded beyond the finance of entrepreneurship to the finance of strategic and emerging industries. A key GGIF was the National Integrated Circuit Industry Investment Fund (known colloquially in China as the Big Fund) with its objective to reduce China's heavy reliance on foreign semiconductor technology. In 2014, the Chinese State Council reiterated its support for measures to promote 'mass entrepreneurship and innovation' and local governments were encouraged to establish their own GGIFs in support of this policy.

As part of the pro-entrepreneurship policy, effort had been put into creating clusters of entrepreneurial activity around the country. By 2010, the startup-VC-IPO conveyor belt familiar in Western economies had taken root in China, leading to the emergence of a thriving startup culture in and around several cities, notably Beijing, Shanghai, Shenzhen and Hangzhou. Engineers with bright ideas in these and other locations were able to access capital to pursue them.

Why did the Chinese construct this edifice of so-called guidance (effectively, investment) funds? The basic reason is that the necessary domestic foundation for venture capital did not exist in the way it does in Western nations. Chinese society lacked long-established, well-endowed and experienced investment institutions (pension funds, insurance companies and foundations of various kinds) to underpin the system. As an example, Chinese insurance companies were not permitted to invest in venture capital until 2010. In contrast with the dearth of institutional investment, no problems were encountered replicating Silicon Valley-style venture capital partnerships. The VC firms that emerged were largely home-grown. One of the earliest was Hillhouse Capital, which started in 2005 with a $20mn investment from the endowment of Yale University where its founder had studied.

The leading American venture capital firms have only had a limited involvement in the Chinese VC boom. The single striking exception is Sequoia Capital, one of the doyens of Silicon Valley and an early investor in Apple, Google and PayPal. Sequoia set up a subsidiary in China in 2005. Sensibly, the partners appointed Neil Shen, a Chinese national and a successful entrepreneur and investment banker, as managing partner. Shen became the driving force behind the explosive growth of Sequoia Capital China. He became well known in Chinese circles as a member of Beijing's top political advisory council. However, amid US–China tensions, in early 2023, his membership was withdrawn. Sequoia Capital China subsequently announced that it would split from its parent company and operate independently as HongShan.

CHINESE TECH IPO

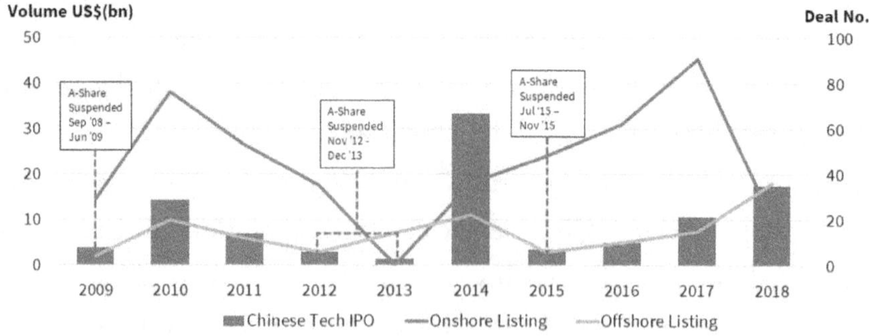

Source: Dealogic, 28 February 2019

Although heavily regulated, the stock market mechanism that evolved after 2010 functioned well enough. As illustrated above, technology IPOs became an established part of the Chinese corporate landscape.

Chinese entrepreneurs, when making the decision to seek venture capital have had the choice between US dollars or Chinese renminbi. As a general rule, raising dollars implied the ultimate goal was a listing in New York or Hong Kong while choosing the domestic currency usually resulted in a listing in Shanghai, Shenzhen or Beijing. In 2022, amid worries about sanctions, caution on the part of international institutional investors and a push by the authorities to exert a firmer grip on the reins of the venture financing system, the balance swung decisively in favour of renminbi. Following a retreat by foreign investors, aggregate funding has fallen sharply from its peak in 2016. New rules in March 2023 discouraged Chinese entrepreneurs from seeking to float on overseas stock exchanges.

To ensure the availability of funding for innovation, the authorities permitted the formation of a dedicated stock market for technology in 2009, Shenzhen's ChiNext market. This was followed in 2019 by Shanghai's more ambitious science and technology-focused STAR Market, with the avowed aim of becoming China's version of Nasdaq, the US tech-heavy securities marketplace. Two years later Xi Jinping, the Chinese president, launched the Beijing Stock Exchange with a brief to support tech startups that might struggle to meet the stricter criteria of larger exchanges. Then, in 2025, the all-powerful regulator, the China Securities Regulatory Commission, relaxed the rules on high-tech companies seeking a listing in Hong Kong, which had enjoyed an IPO boom after 2010, encouraging firms to use this route in place of an IPO on a foreign market. Consequently, China now has a plurality of tech-focused stock markets.

GGIFs, LGFVs AND CVCs

In 2014, when the government decided more money should be directed towards 'mass entrepreneurship and innovation', Chinese institutional investors were few in number, inexperienced and ultimately state controlled. The CCP apparatchiks who ran them were – and are – extremely risk averse, not least because of the political repercussions if they got it wrong. As such, early-stage venture capital had zero appeal to them. The venture capital infrastructure was in place, but something was needed to ensure a plentiful supply of capital to these partnerships. Hence the need for a re-energised GGIF structure.

Government-guided investment funds are there to supply capital. Many were set up – there are around 1,800 now. They are essentially passive investors in the companies they finance, usually standing behind VC firms in the conventional manner. A fund is an entity formed by, or at the behest of, central, provincial or local government. The sponsor sets up the fund and then seeks equity from private sources. In practice, the majority of these co-investors are state-run banks or state-owned enterprises. Consequently, their designation as 'public-private' funds is misleading – they are more public than private. The graph below shows the major impact guidance funds had in 2015 and 2016. (It covers private equity as a whole, not just early-stage investment.) By 2017 private investors had taken up the running from GGIFs, as was intended.

As well as providing finance, GGIFs are there to ensure that the investments they make are consistent with central government's strategic policy goals. Financially, they have undoubtedly had a significant multiplier effect by conferring an official stamp of approval on the early-stage funds of the VC firms they chose to invest in. For hesitant executives in state-controlled organisations, it was as if the ultimate boss located somewhere in Beijing had hoisted a 'Go ahead, it's OK to

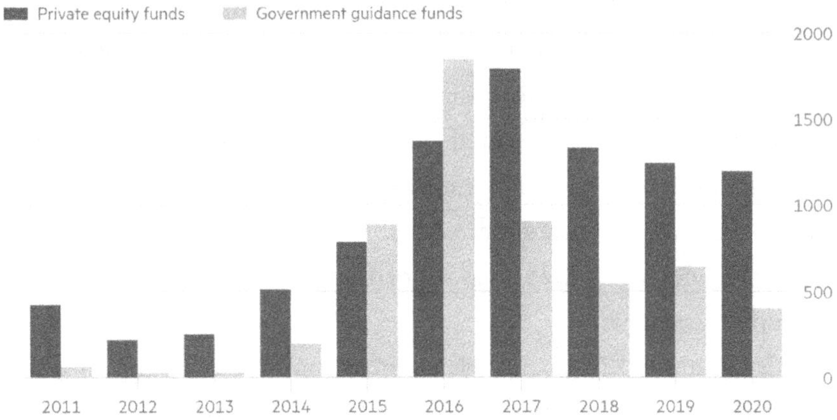

Source: Zero2IPO in Asian Review of Political Economy, 9 August 2024

invest in early-stage funds' sign. Thus reassured, other investors – local governments, banks, insurance companies, pension funds, private individuals – followed suit.

Central government has encouraged local governments to borrow to fund infrastructure projects via yet another bewildering set of initials: LGFVs, standing for Local Government Financing Vehicles. The result was a spending spree that spilled over from infrastructure to investment in local businesses, often with little concern for commercial viability. By 2020 it was recognised that some LGFVs were in trouble. They had taken on too much debt and, under pressure from Beijing, have been forced to rein in their activities.

The infusion of capital from domestic institutions has been accompanied by a rise in corporate venture capital (CVC is the shorthand). Echoing practice in the West, successful Chinese companies invested their own capital in promising startups. The motivation, as elsewhere, was twofold: to identify ideas and technologies that could be of practical benefit to the business down the line; and to make money from the investment via an exit. In an ideal world, the company achieves both of these objectives. The domestic internet giants, Baidu, Alibaba and

Tencent were at the forefront of this trend, investing in companies well beyond e-commerce. Other successful companies, such as Huawei and Xiaomi, followed suit by forming in-house venture capital units. Since 2015 Chinese CVC investment has accelerated to become a major force, rivalling conventional venture capital as a source of funding. Major Western CVCs are still active. A good example is Qualcomm Ventures, which, in 2025, had a portfolio of more than 50 Chinese startup investments.

International institutional investors, notably US pension funds and endowments, are now markedly less important to the Chinese tech startup ecosystem, their contribution having taken a deep dive in 2022. According to S&P Global Market Intelligence, the international investor inflow dropped from $28bn in 2021 to $9bn in the following year. It has continued to fall. Nor have domestic institutional investors in VC been prepared to take up the slack in the face of a declining economy and uncertainties following political intervention in business. By 2025, China had become heavily reliant on public money to underpin its tech startup ecosystem.

To sum up: China has been successful since 2014 in creating a vibrant startup culture with funding to sustain it, although the reduction in international commitments, and the reduced funding capacity of LGFVs, leaves a gap to be filled. Consequently, this Chinese variant of the Western venture capital model is not without its issues, above all the ever-present prospect of capricious state intervention. After 2015, several Chinese VC-backed and newly listed companies came to play a role in CML's progress, as suppliers, collaborators and customers.

13.

Learning to fly with HTI

A CHEQUERED HISTORY

The aerial analogy in the title of this chapter is apposite. Hutchinson Technology Inc.'s mission was to get SMA actuator production off the ground to a 'low cruising altitude' (one million units a month) as speedily as possible. As HTI was a publicly listed company – and had been for a long time – it is possible (and useful) to view the sequence of events as the relationship developed from 2013 onwards through the twin lenses of CML and HTI simultaneously.

To understand HTI's reasons for linking up with CML it is important to appreciate the company's history. The deal with CML was not the first time that the company had tried to diversify from its existing product for Hard Disk Drives. In 1996 HTI set up a division to develop and sell medical instrumentation, drawing upon the company's expertise in precision measurement. The division remained small and never made a profit. In November 2012 – just about the time CML first came into contact with HTI – Timothy Stabosz, who had become the second largest shareholder in HTI, wrote to the board of directors calling for the division to be sold as it had 'lost a staggering $154m from operations, in aggregate, since 2000'. He went on to say that:

> '... the last several years have been "incredibly painful for Hutchinson Technology's shareholders" who he claims have watched the company's market share drop from more than 70 percent to about 20 percent. And if the company does not find itself "on the road to profitability" by the fourth quarter of its 2013 fiscal year, it should consider selling itself to a competitor or a disk drive maker ...' ('Major Hutchinson Tech Investor asks Co. to Sell Division', Banking+Finance, 23 November 2012)

Within a year of this broadside, the division had been substantially wound down.

The fact is that, for long-term investors in HTI, it had been a 28-year wild ride. Hutchinson Technology Inc. went public on Nasdaq at $10 a share. The stock price reached a peak of $45 in the dot com boom and then collapsed spectacularly in 2008 to a few dollars, from which it never recovered. The company was forced to raise additional debt, totalling $57mn over the three years to 2013.

Rick Penn had been promoted to CEO in September 2012 with a brief to get the suspension assemblies business back into solid profit. Given the cyclical nature of the HDD industry, this was a tall order. Production of suspension assemblies is a capital-intensive business, requiring constant investment in new and upgraded equipment. In the background, investors were worried by the steady incursion of solid-state drives (SSDs) into the computer storage market.

What lay behind the pressures on Hutchinson? In essence, the problem was that HTI was a small fish swimming in a tank of oligopolistic leviathans. There were few buyers of suspension assemblies, and it was competing with rivals considerably larger and more vertically integrated. There were really only three customers for HDD assemblies: Seagate Technology, Toshiba and Western Digital. Japanese companies had mastered the art of producing components for HDDs and it remained a stronghold for them. Most troubling from HTI's point of view was that one of these competitors was owned by TDK. As we have seen previously in this narrative, TDK's tentacles extend far and wide in electronic and electro-mechanical components. Being vertically integrated, it could undercut HTI by offering a package to HDD customers.

HTI's share of the HDD suspension assemblies' market hit 20 per cent in 2013, having been 50 per cent a few years back. The need for an additional product to defray the overheads was urgent. HTI expressed confidence that it could meet the volume demands that CML envisaged, confidence grounded in the company's experience of high-volume production. After all, even in its reduced state, HTI was making and shipping 35mn units a month.

GEARING UP FOR SMA

At the beginning of 2014, CML was telling its shareholders that HTI was confident it could achieve a run rate of one million units a month by mid-year, in time to meet the expected initial rush of OIS-enabled models. But, by the end of the period, HTI was only making 100,000 actuators a month.

HTI announced its involvement with CML to the US stock market on 1 August 2014. The stock price reacted favourably to the news, climbing 30 per cent on the day. At the analyst conference call, Rick Penn talked about 'ramping up volume later this summer' from an OIS automated production line that had already been installed. He also revealed that 'The initial SMA OIS product will be used on a smartphone for the Chinese market that will be released in the fall of 2014.'

Towards the end of the year the reality was less rosy. Final yields were much too low – running at 35 per cent – but HTI's engineering team remained confident they could raise them to the 95 per cent they were achieving on suspension assemblies. The output target had been downgraded to 600,000 units a month by December. In truth, there was no immediate demand pressure, as the design wins obtained by CML via Foxconn were taking time to turn into production orders. Publicly HTI continued to exude confidence.

HTI was making enough actuators for the modest volumes required by the InFocus M530 when it went on sale in January 2015. The pressure on the original business had not abated and HTI was bleeding cash. On 31 March 2015, the company reported it had debts of $128mn, of which $76mn was due to be repaid in January 2017. Uncomfortably close. Late in 2015, the new design Gemini samples were sent to New Shicoh and TDK for evaluation. TDK's major customer was Huawei, known to CML from its initial contact via Foxconn. CML was keen for HTI to manufacture an initial quantity to prove the process and create stock in anticipation of orders, but HTI was reluctant. Such was the pressure on working capital that building for inventory was discouraged.

Then on 2 November 2015 came the surprise stock exchange announcement of an agreed merger between HTI and TDK. The takeover – 'rescue' would have been a more accurate description – was primarily driven by TDK's desire to consolidate the HDD component business. HTI's foray into SMA actuators was simply a nice add-on for TDK. Investors in HTI breathed a sigh of relief at the all-cash offer and the shares doubled on the day to $3.35. In the subsequent quarterly results announcement Rick Penn remained upbeat on the prospects for SMA OIS, saying that:

> 'We have shipped prototypes of our Gemini OIS actuator to lensholder assembly manufacturers and currently expect some smartphone makers to begin receiving prototype camera modules that use our new OIS actuator by the end of this calendar year.' (Hutchinson Technology annual report for fiscal 2015)

From CML's perspective, the effort to get HTI up to speed as a reliable source of actuators had been a deeply frustrating experience. Translating Hutchinson's undoubted acumen in the mass manufacture of suspension assemblies to the mass manufacture of an unfamiliar component proved to be much harder than the engineers at HTI had anticipated. Recognising this, CML intensified its efforts to find an alternative production source. As part of this process CML took a hard look at the restrictions on creating a second source in the agreement with HTI. More generally, TDK's bid to absorb HTI appeared to be good news given CML's existing links with TDK. Assuredly, HTI's production capabilities would be boosted by the engineering resource that TDK could bring to bear.

Meanwhile, discussions with Samsung were continuing. The year to December 2014 had been tough for Samsung Mobile. Low-cost competition from up-and-coming Chinese producers had sliced into the Mobile Division's revenues and profits. By its own admission (in the company's annual Business Report) its world market share in smartphones had fallen to 22.4 per cent in 2014 from 26.8 per cent the year before. Samsung was looking for anything that would reduce cost and differentiate their phones. SMA OIS was one such option. What happened to these discussions is picked up in Chapter 17. Before that, we need to look at CML's finances, how Voice Coil Motors resisted SMA actuators, and understand the characteristics and foibles of the leading smartphone manufacturers

14.

Keeping the show on the road 1: management, finances and strategy

BOARD AND MANAGEMENT CHANGES

The purpose of this and subsequent chapters is not to tabulate CML's financial history or record in detail the changes in personnel that took place over the years. The intention is rather to identify the key events along the way.

Stewart Newton first became a Cambridge Mechatronics (or 1...Limited, as it then was) shareholder in 2002 as part of a funding round. Philip Bradley, who had become a director of CML, urged his contacts to invest. The fact that he had committed a large sum himself – 'put his money where his mouth is' in the classic City phrase – acted as an incentive to others. Stewart Newton was an indirect contact – a friend of a friend. A chartered accountant by training, he had built an extremely successful investment management business from scratch. In 1998 he accepted a bid for 75 per cent of Newton Investment Management from Mellon Financial Corporation. The remaining quarter was purchased by Mellon over the subsequent four years through staged payments.

By May 2002, after 25 years at the helm, he was able to withdraw completely from Newton Investment Management and became free to concentrate on other interests. The total consideration has never been disclosed but it is believed to have run into tens of millions of pounds. It is worth reiterating the absolutely critical role that Stewart has played in the survival of CML. Without his unwavering support, the company would have vanished from the scene years ago. The coincidence of Stewart's sale and 1...Limited's increased requirement for cash in the early 2000s was indeed fortuitous.

It happened that Newton Investment Management employed specialists in various fields to assist in making investment decisions: 'in-house investment analysts' in City terminology. In April 2000, Newton announced a new fund to 'take advantage of global opportunities in technology'. The manager of Newton Global Innovation, as the new fund was called, would be assisted by Newton's global technology and telecommunications specialists.

One of these was Dr Fiorangelo Salvatorelli, senior global research analyst and director. He was exceptionally well qualified for this role, with an impressive science and business pedigree. Educated originally in Venezuela, he obtained his doctorate in Engineering Science from the University of Oxford, became a lecturer there and then spent seven years as a management consultant with McKinsey and Company. He joined Newton Investment Management in 1995 and left in 2004. When contemplating an investment in CML, Stewart naturally consulted Fiorangelo. With his deep knowledge of the electronics sector, and having taken a look at the company, Fiorangelo was enthusiastic and, indeed, invested on his own behalf. He remained close to Stewart in regard to CML and continued to advise him. Fiorangelo was subsequently appointed an advisor to the board of CML.

As CML geared up to produce the Digital Sound Projector and expand research into actuators, 2002 turned out to be a busy year for fund-raising. Nearly £4mn was brought into the company in two stages. By the end of this process, Tony Hooley's stake had been diluted to a bit less than a quarter of the equity while two unidentified private shareholders between them controlled another quarter. It is assumed that Stewart Newton was one of these two unidentified investors. His investment in CML was held personally and also in Sussex Research Limited, the Newton family investment vehicle.

Dr Martin Knight was invited to join the board in June 2003. This turned out to be a highly significant appointment. Due to his long association with Imperial College, London, he was attuned to the problems of financing science-based startups. Martin had been a member of Imperial's Governing Council since 1994 and chairman of its Finance and Investments committees. He has a doctorate from Oxford University and, after a career in investment banking, was appointed chief operating and finance officer of Imperial College, London, in October 2004. Additionally – and of more direct relevance to CML – the year before he had become chair of Imperial Innovations, the College's commercialisation and investment arm, a post he held until 2016.

By May 2004 Stewart Newton had evidently become convinced that CML had a bright future. The licence agreement with Yamaha had just been signed but CML wanted to push on with mobile camera production designs and was running out of cash. To plug the gap, CML proposed to issue shares valued at £1mn to existing shareholders. Not all were eager to put more money in so, to ensure the company would receive the full amount, Stewart committed to make up any shortfall. The Information Memorandum said:

> 'The round has been underwritten by Sussex Research Limited, a company connected to Stewart Newton. Depending on the take-up from other existing shareholders, Sussex Research Limited

(together with Stewart Newton) will hold between 20.07% and 25.52% of the issued shares at the end of this funding round.'

In return for this guarantee, warrants were issued to Stewart for the purchase of CML shares at a specified price. Stewart's direct support for this cash-raising exercise set a precedent.

The directors of CML elected Martin Knight as non-executive chairman in 2004, replacing David Gillett. It was not long before the board had to contend with the Helimorph debacle. By the end of 2006 Mark Shepherd, CML's CEO since 2001, had left. Tony Hooley returned to his old position as CEO and Andy Osmant, CFO (chief financial officer), took on further commercial responsibilities.

Inevitably, more money was required. The board and investors looked to Tony Hooley for reassurance that there was a future for the DSP and for the SMA actuator technology still under development. Tony provided that reassurance by investing a considerable sum in the new funding round and heading up the presentations to investors. Once again, Sussex Research, together with one other large shareholder, underwrote the issue of new shares. In the following few years, the company was kept afloat by regular issues of convertible and non-convertible loan notes underwritten by Sussex Research. These loans were subsequently repaid or converted into equity.

STRATEGY AND SHAREHOLDERS

Was there a point in the CML timeline when the directors stood back and considered whether to pursue the actuator opportunity? Indeed there was, in November 2007. Further finance was urgently required, and it was not at all clear whether it would be forthcoming. The board presented shareholders with three options:

1. To abandon the actuator business and downsize to the audio business.
2. To sell the audio business and reinvest the proceeds in the actuator business.
3. To retain the company as a single entity and raise funds to pursue actuator development.

In the event, option three prevailed though, with the sale of audio to Yamaha in 2013, CML eventually went down the path of the second option. Then again, in July 2015, the directors of CML engaged in another round of soul-searching, positing four longer-term options in a shareholder presentation. The decision was again made – in the wording of the presentation – to 'Go for it'.

In January 2008 Simon Calder was brought in to be the new CEO of CML, allowing Tony to step down from what he always considered to be an interim role. By this time, major restructuring had cut costs and losses were reducing. As a consequence of his underwriting efforts, Stewart Newton was identified in the 2008 annual accounts for the first time as 'the ultimate controlling party' (over 50 per cent). CML made a profit in 2009, thanks to the continued suppression of costs and bumper fee revenue from Yamaha. Headcount had been cut from 39 in 2007 to 18 in 2009. In September 2011 Stewart Newton joined the board.

At the end of 2012 shareholders were informed that Simon Calder had resigned and Martin Knight had moved to take a much more 'hands on' role in the running of the company as executive chairman. There would be no chief executive: Andy Osmant became managing director in December 2013. By this juncture, Martin had amply demonstrated his belief in the future of CML by buying heavily into the equity. Licence fees from Yamaha and further underwritten loans continued to fund the company. In June 2014, Stewart Newton owned 64 per cent of the equity. A presentation to shareholders showed the ownership of CML to be:

S Newton	51%
Sussex Research	13%
M Knight	5%
A Hooley	3%
Staff share options	8%
Other shareholders	20%

Source: Cambridge Mechatronics Ltd (CML)

On the basis of the (low) price of this funding round, CML was valued at a modest £22mn. However, in mid-2014 there was hope in the air. The company was out of the audio business. It may have failed to secure business with a Tier 1 smartphone vendor, but China was opening up fast. CML had secured a manufacturing agreement with Hutchinson Technology Inc.. Lines of communication with Samsung Mobile were still open. In August 2014, CML recruited Warren East, whose 12-year tenure as CEO of Arm had been manifestly successful, as an adviser to the board. Arm's unique status as the designer of the architecture on which every smartphone SoC is based meant that he brought with him an unrivalled knowledge of the mobile phone market.

15.

VCM fights back

VCMS RULE

It will be clear by now that the incumbent technology of Voice Coil Motor was the big stumbling block to the adoption of SMA autofocus actuators in smartphones in the period 2008 to 2012. Even with the move to OIS, VCMs retained their grip. How this was achieved is the subject of this chapter.

An outsider looking at the mobile phone industry in 2005 would surely have come to the conclusion that the VCM as a means of moving lenses was 'over the hill' and that, in short order, it would be displaced. The Voice Coil Motor had been invented in the mid-nineteenth century, but it was well established and familiar to smartphone camera designers. So they stuck with it despite its shortcomings. Chief among these was its high power consumption. Designers were also aware that the VCM's ability to provide motive force to lenses was limited. And it was thought that the magnetic properties of VCMs would increasingly inhibit handset design.

Voice Coil Motors are based on a copper coil energised by an electric current and placed inside a magnetic field provided by magnets. Applying a voltage causes the motor to move a few tenths of a millimetre in one direction or, by reversing the polarity of the applied voltage, the opposite direction. VCM actuators are constructed from a variety of metal parts and are therefore complicated to assemble. Coils need to be wound precisely to achieve the required characteristics for the application at hand. The design also involves metal housings and metal springs. There are at least ten individual parts.

Initially, there appeared to be limited scope for making actuators small enough to meet smartphone requirements. VCM makers, however, rose to the challenge. The threat from SMA – and other budding actuator technologies – spurred the VCM world into action. CML was far from alone in viewing smartphone camera actuators as a major growth opportunity. A variety of companies from diverse technological backgrounds tried to develop robust, low-cost actuators for phones. It proved to be far harder than envisaged. My calculation of the total commitment to finding a better way of controlling lenses over a period of more than 20 years is close to one billion dollars. Who these competitors were, the technology on which they based their challenge and what happened to each of them is covered in Appendix 3. To explore the appendix, scan the QR code or visit my book website at www.downtothewirecont.com

Consumers wanted slimmer phones and manufacturers were obliged to respond. When phones were thicker than 10mm, a chunky actuator was not an issue. However, in the mid-2010s, average handset thickness dropped below 10mm. Today's phones are commonly between 6 and 8mm in depth. Impressive as it is, the continued compression of camera modules has not been able to keep pace with the quest for thinness, and cameras often protrude (the well-known bump), preventing the phone from lying flat on a table.

How did VCM see off the challenge from various directions? In essence, consistent, incremental improvements in key constituent components, together with design and production engineering innovation, produced, over time, a step function uplift in VCM performance and enabled it to stay competitive. There is, however, a caveat. SMA and its competitor technologies were not aiming to supplant the entire population of VCM actuators in mobile phone cameras. Half the handset market contains low-spec and low-priced cameras. These devices use AF actuators that are made very cheaply by small Chinese enterprises, often subsidised by the Local Government Financing Vehicles (LGFVs) described in Chapter 12. Given the miniscule – or even negative – margins earned by these tiny firms, there was no point in trying to compete.

Rather, the objective of the challengers was to replace the VCM actuators used in OIS and AF actuators at the higher-priced, quality end of the market. In the mid-2010s, the premium segment represented 15–20 per cent of the global smartphone market by numbers. Fortunately for CML, premium phones have accounted for a steadily growing proportion of the market since then. Defined as a phone with a wholesale price exceeding $600, these now represent a quarter of the market.

THE MAGIC OF MAGNETS

Few realise it but magnets are a minor miracle of modern technology. They lie at the heart of every Voice Coil Motor. Specifically, VCMs use permanent magnets – those that retain their magnetic properties once magnetised – made from an alloy of several rare earth elements (REEs). One of these in particular, Neodymium, is used in quantity while others play a minor, though critical, role. Mobile phones use magnets in several areas, not just the lens-moving motor. Smartphone loudspeakers, for instance, contain between three and five permanent magnets per speaker. The number of magnets per phone adds up quickly, as per the image opposite.

Source: Zhejiang Innuovo Magnetics Co.

Once again with due deference to Jane Austen, it is a truth universally acknowledged – by those whose job it is to know – that without the astonishing progress made in the performance, size and weight of permanent magnets, the slim, highly capable smartphones of today would simply not be possible. A VCM-based OIS camera system can use as many as four permanent magnets in each actuator. This has potential downsides. It can give rise to magnetic interference, both within the camera module, between camera modules and with other magnetic-based components. Handset designers need to ensure that magnet-containing components are not too close to each other. SMA has, of course, no such constraints.

Rare earth elements are not actually that rare but are extremely difficult to mine and refine. Neodymium combined with iron and boron (and some additional REEs) produces a very strong, small and light permanent magnet, commonly referred to as NdFeB. NdFeB magnets deliver the highest performance in the smallest volume of material and have been instrumental in the progressive miniaturisation of electronics. The largest single source of demand for NdFeB magnets is in the motors of electric and hybrid vehicles. They are also used in electric motors, MRI scanners, electric generators for wind turbines and Hard Disk Drives.

Economically exploitable deposits of rare earths are relatively sparse. Since the early 1990s, China has become the world's primary source. Rare earth mining and downstream processing in China has benefited from lower labour and energy costs and laxer environmental standards. Acids and chemicals used to extract rare earths leave a toxic legacy in terms of soil, air and water pollution. Some raw material is produced in countries outside China but most of that concentrate is shipped to China for processing and refining, as China possesses nearly all the large-scale facilities to do this currently. An estimated 90 per cent of NdFeB magnets are manufactured in China, with obvious geopolitical implications.

In recent years, improved production techniques have contained the costs of Neodymium-based magnets. More to the point, advances in processing have also led to an improvement in the properties of these magnets. Permanent magnets account for no more than 10 per cent of the cost of making a VCM but, in terms of performance, their contribution is vital. Essentially, advances in magnets have underpinned progress in VCMs.

VCM CHAMPIONS

Voice Coil Motors still mostly rely on coil winding, a labour-intensive task. Better design of VCMs enabled automation of the production process. Automation drove down cost and simultaneously improved quality. Operators sit at lines of software-controlled machines, performing several functions previously carried out manually. Chapter 6 described how, with one outstanding exception (Sony), Japanese digital camera and optical component producers singularly failed to translate the solid position they held in dumbphones into a comparable share of the smartphone market. There is, however, an area where Japanese firms have made their mark: VCM camera actuators. Three, large, established Japanese electronic component companies, Alps Electric, Mitsumi and TDK, secured their position in VCM actuators early in the evolution of the camera phone.

Japanese industry has a strong tradition of production engineering, and Alps Electric, Mitsumi and TDK pioneered automated coil winding machinery. Alps has been in the autofocus VCM actuator business since 2010 and in mass production since 2015. The company exemplifies the emphasis on production engineering that characterises all three Japanese actuator makers. As the page on production technology in the Alps Alpine website puts it:

> 'The key to making progress with automation and process design is having production equipment in house ... Alps Alpine set to work on developing its own production equipment from an early stage, and since that time has continued to conduct updates while quickly incorporating the latest technologies and trends of the day. This attitude of independently making our own equipment has been carried on for a long time since the founding of the Company, and it has become one of our strengths.' (Alps Alpine website, Production Technology, *The Optimization of our Production System Is Made Possible by our Production Facilities for In-house Manufacturing*)

In 2008 Apple needed an autofocus actuator supplier for the iPhone and established a relationship with Mitsumi. Apple then wanted a second source and, to this end, insisted that the VCM production technology that Mitsumi had developed at Apple's behest be handed over to Alps. The dual source relationship that developed was subsequently extended to OIS and continues to this day. It has been good for both companies and for Apple.

Largely due to the secure position they retain with Apple, Alps (now Alps Alpine) and Mitsumi (now MinebeaMitsumi) still figure prominently in the VCM actuator rankings. However, competition from Chinese and South Korean manufacturers has intensified and eroded their market share. As explained previously, there exists a long tail of small Chinese producers, but these have been joined by several competent, mid-size operators. One of these, New Shicoh, has been mentioned previously in regard to the SMA licence it negotiated with CML in 2014. As we pursue this narrative, we will encounter several more Chinese VCM manufacturers in the context of CML's desire to spread the gospel of SMA OIS.

Estimated VCM camera actuator suppliers market shares 2024

RANK	COMPANY (HQ LOCATION)	ESTIMATED MARKET SHARE BY REVENUE
1	Alps Alpine (Japan)	19%
2	MinebeaMitsumi (Japan)	11%
3	ZET (China)	10%
4	SEMCO (South Korea)	10%
5	New Shicoh (China)	10%
6	TDK (Japan)	9%
7	Hozel (China)	8%
8	Jahwa Electronics (South Korea)	8%
9	Billu (China)	4%
10	L G Innotek (South Korea)	3%
	Others	8%
Total		**100%**

Source: Author, compiled from technology research consultancy reports

ALPS, MITSUMI AND TDK

Alps, Mitsumi and TDK run on parallel tracks as suppliers of VCM actuators but their strategies for this product have differed markedly, with down-the-line implications for CML. At Alps Alpine not only does the iPhone contract drive the VCM operation, but this business is also central to the future of the company, accounting for an estimated 15 per cent of revenues. In its public statements Alps is – as are all Apple suppliers – circumspect about the destination of its actuators. The major customer is never named. For example, the report of the results for the first half of 2021/22 commented on the demand from 'North American manufacturers set to launch new-model smartphones'. This is obfuscation: there is only one North American manufacturer!

Alps Alpine's sales have declined since 2017, and the operating income margin has fallen substantially. In essence, OIS actuators have performed well since 2021 in a sea of mediocre or negative returns elsewhere. Taking market share from Mitsumi, the company has reinforced its position as the global leader in smartphone camera actuators. In terms of its financial record and its share price performance, Alps has long been a laggard. Quite why it has languished is unclear. Alps Alpine is a company that has still to convince investors it can grow consistently and profitably.

In the actuator marketplace, Mitsumi has steadily lost ground to Alps. Mitsumi decided in 2015 it needed to bulk up and merged with a ball-bearing manufacturer. The merged company adopted a cumbersome new name: MinebeaMitsumi. On the face of it, adding ball bearings to electronic components seems an odd amalgam but there are considerable synergies in the two companies' technological skillsets. Camera actuators – a strategic pillar a decade ago – have now become peripheral, accounting for an estimated 4 per cent of group sales. However, MinebeaMitsumi, in contrast to Alps Alpine, has grown revenues and profits steadily. The relative share price performances of the two companies reflect this stark reality.

TDK, the third Japanese actuator maker, is a different kind of Japanese corporation from the other two. It is larger and more international in its outlook. From its original base in magnetics and ceramics, it now offers a highly diverse range of electronic components. TDK has long had a policy of adding to its product portfolio by acquisition. These have tended to be small to medium-sized US and European companies with a particular specialism. The largest of these was the American gyro specialist, InvenSense (referred to in Chapter 12 in the context of OIS) in December 2016. From the start, TDK focused its actuator business on the Chinese market, supplying the likes of Huawei. But, without the steady baseload provided by the iPhone, TDK has never achieved the same level of success as its two fellow Japanese camera actuator suppliers.

What of actuators in TDK? In the corporate reorganisation that followed the acquisition of InvenSense, TDK was unsure what to do with the SMA actuator business it had inherited from the HDD-driven acquisition of Hutchinson Technology Inc.. There was no mention of camera actuators in the 2018 annual report. Then suddenly, in subsequent annual reports, OIS actuators popped up in the residual category of 'Other' for segmental reporting purposes, lumped in with 'Load ports, flip-chip bonders, flash memory application devices, anechoic chambers, etc.'. 'Other' made no discernible profit contribution to TDK. Borrowing a railway analogy, it certainly looks as if actuators were, from 2019, shunted into a siding. Proof of this came in April 2025 when TDK announced the sale of its actuator business to the Chinese camera module integrator, Q Technology, which has been listed on the Hong Kong Stock Exchange since 2014.

AN EXTENDED LIFE

It will be apparent by now that smartphone vendors, led by Apple and Samsung, had no burning desire to move away from the venerable camera actuator technology that had served them well. VCM was proven and freely available from several suppliers with established low-cost production facilities. Rather than taking a risk on something new, it was surely better to devote R&D resources to extending this trusted and familiar technology as far as it would go. So this is what they did.

Having tried – and failed – to find an alternative, Apple made a decision to try to refine and extend VCM. The preamble to an Apple patent filing in 2021 is revealing:

> 'As the cameras become more complicated, they become bigger and heavier, and more powerful actuators are needed to deliver the required motive force. Traditional VCM actuators can be power-limited due to the size and energy source limitations associated with mobile devices. Thus, it is desirable to have VCM actuators with more efficient architectures to boost output power.' ('US Patent Application for Multi-Coil Voice Coil Motor Architecture', Justia Patents, 2 April 2021)

Apple has therefore continued to push the VCM envelope in search of enhanced performance and additional efficiencies.

SEMCO, the Samsung affiliate, has followed a similar path, devoting R&D resources to advance the performance of VCM OIS actuators. Building on research published in a Korean academic paper in 2010, SEMCO pursued the concept of using ball bearings instead of springs. The resulting 'ball guide' actuator is a good illustration of the way in which it has been possible to upgrade and extend VCM technology. Samsung claims that the ball guide method – where the lens moves as the ball rolls – is advantageous in operational speed but, more importantly, it consumes less battery power and is capable of moving heavier lenses than the standard VCM actuator. Samsung, through SEMCO, continued to invest in the development of the ball guide actuator during the 2010s. We discover what happened to Samsung's proprietary actuator technology in Chapter 23.

COMPETITION FROM CHINA

By 2015, the three Japanese actuator leviathans were facing vigorous competition from Chinese entities – often newly minted and funded by Local Government Financing Vehicles – in AF actuators. These producers benefited from subsidies via the cheap loans that LGFVs were able to access, allowing them to operate at low or negative margins. As the price of AF actuators declined, the Japanese companies retreated from AF to concentrate on the more complex and more expensive OIS actuators (for premium phones), leveraging their well-honed mass production skills. We saw in Chapter 12 that CML had linked up with New Shicoh, a prominent Chinese VCM actuator producer. Other leading Chinese firms in 2015, apart from New Shicoh, were Shanghai B.L Electronics Co. (Billu) and Zhonglan Electronics (ZET), both of which were to play a significant part in the CML story as it unfolded.

THE SAILING SHIP EFFECT

To sum up the theme of this chapter: VCM actuators were able to resist the challenge from alternative lens-moving technologies through a quantum leap in magnet performance accompanied by a relentless pursuit of techniques to improve design and reduce the cost of production, in large part by replacing people with machines. Japanese production acumen has allowed them to retain much of the VCM OIS business in the face of Chinese competition, but Chinese manufacturers have come to dominate the lower-end, lower-cost VCM AF market.

To overcome the sheer conservatism of smartphone customers, a new technology needed, first, to demonstrate *major* advantages over the one currently in use and, second, a track record of units actually in the market. SMA was unable to meet these criteria – but then nor did any of the other companies striving to discover and commercialise a better way of moving lenses in smartphone cameras (Appendix 3).

This phenomenon of an established technology surviving for a long time after a superior solution emerges is well recognised by academics who study innovation. It is known as the 'sailing ship effect', referring to a situation in which an old technology is re-invigorated when faced with the risk of being replaced. Following the introduction of steamships in the 1850s, sailing ships continued to be an economical way of transporting bulk cargos, such as lumber or grain, on long voyages as late as the 1940s. It is surely a supreme irony that, in the 1870s, they were extensively employed to transport coal to far-flung bunkering stations. Several innovations increased the speed of sail. Between 1870 and 1900 windjammers or tall ships, the largest merchant sailing ships, were built with iron hulls and steel masts, replacing wood. Hull shapes evolved from being relatively short and blunt to being longer and finer. Onboard labour-saving machinery substantially reduced crew requirements, so bearing down further on costs.

Consequently, even in 1880, most of the goods transported across oceans were still going by sail. The changeover from sail to steam did not occur on most long-haul routes until the very end of the nineteenth century. Other examples of the sailing ship effect include the response of gas lighting in the 1880s to the arrival of incandescent lamps. In more recent times, during the 1980s, carburettors dramatically improved their fuel efficiency when faced with Electronic Fuel Injection. Voice Coil Motors fit well into this paradigm.

16.

Smartphone makers: the ones that matter

SETTING THE SCENE

For CML to succeed, on the basis of the royalty-based licensing model it had adopted, one thing was imperative: actuator production volume. This entailed maintaining a focus on getting business from the largest smartphone companies, specifically those with an emphasis on selling the premium phones that increasingly featured OIS. Securing business from minor players would not, even in aggregate, be enough. The five leading companies in smartphones are very different from each other. Any marketing strategy has to take account of their characteristics and foibles. In this chapter we explain the differences.

An IP licence model is very simple in essence: the company sells a manufacturing organisation the right to make the product under licence. The licensor additionally contracts to provide such assistance and know-how as may be necessary to get the licensee up and running. The licensor's financial return comes in the form of two income streams: up-front and stage cash payments, followed by a royalty levied on its licensees on every device sold. The initial cash inflow offsets some of the licensor's ongoing operating costs, primarily the employment of highly skilled engineers to maintain and advance the firm's technological base. This annual operating cost is largely fixed. Naturally, there will be some increase each year due to inflation and any expansion in employment.

The second income stream, the royalty per device sold, is by far the more important. To succeed, the company needs its manufacturing licensees to sell their output to customers who are able to buy consistently in large volume. The table below illustrates the point (ignoring, of course, any advance payments). Once a high and consistent annual volume is achieved, the company becomes profitable. And, if annual volumes increase, profit rises exponentially as costs are essentially fixed. The problem is, of course, getting to the take-off point.

Annual operating cost	£10mn	£10mn	£10mn	£10mn
Annual volume	50m units	100m units	150m units	200m units
Royalty per device sold	10p	10p	10p	10p
Annual revenue	£5mn	£10mn	£15mn	£20mn
Profit (Loss)	(£5mn)	Breakeven	£5mn	£10mn

In the smartphone industry the only way to reach the take-off point is by persuading one – or preferably more – of the leading producers to adopt your technology. For a budding supplier, a design win at Samsung or Apple is the equivalent of hitting the smartphone jackpot.

An obvious question arises from the above calculation: the breakeven point would be reached so much sooner if the royalty rate were, say, 20p per device rather than 10p. Beyond smartphones, there are products and customers that offer a higher royalty rate (as explained in Appendix 3, some companies in direct competition with CML have gone down this path). The offset, though, is that, outside smartphones, the volumes – and therefore revenues – are commensurately small. That is the strategic dilemma facing licensors. CML consciously stuck with the 'go for volume' approach.

Why is the royalty rate so low? The short answer is overbearing end-customers. Such is the pressure exerted by smartphone producers on the price of components that manufacturing licensees are left with wafer-thin margins, leaving little for the licensor. Contrast the situation at Dolby Laboratories. It records a respectable margin as a licensor of electronic noise reduction systems because it sells to a diffuse set of industries and customers. Arm, a manifestly successful licensor of technology for smartphones, equally illustrates the point. Arm licenses SoC suppliers, such as Apple and Qualcomm, on the basis of the value of the chip they sell. The royalty rate for its IP reportedly works out at well under a US dollar per unit. This is parallel to how CML operates. At the time of its IPO in 2023, Arm allowed its frustration at what it considered to be the inadequate return from these arrangements to become public. Like CML, Arm is confronted by powerful end customers who drive a hard bargain. How much harder then is it for a small entity like CML to extract an appropriate return for its intellectual endeavours.

The only smartphone firms capable of providing the sort of volumes CML required to become viable are the five profiled below: Samsung, Apple, BBK Electronics, Xiaomi and Huawei.

SKINS AND THINGS

As we observed previously, the purest, most basic version of Android is known colloquially as AOSP, which refers to Android Open-Source Project. Google's annual software updates apply to this proprietary version of Android. Stock Android, as it is often called, refers to the base version of the Android operating system, developed and maintained by Google. Most smartphone manufacturers use stock Android as a foundation, adding their own features. Stock Android includes popular Google Apps, such as Google Mail and Google Maps, and includes access to the all-important Google Play Store. This collection of apps is usually referred to as Google Mobile Services, or GMS. As one would expect, GMS is deeply integrated into the Android operating system.

Each individual smartphone vendor is free to customise its offering by adding what the industry calls a 'skin' on top of stock Android. As with human beings, some skins are thick and others thin. Motorola offers a skin that is close to stock Android. At the other end of the spectrum, Samsung's skin (One UI) offers numerous extra features and customisation options. Xiaomi's HyperOS is an equally thick skin. The fact that Google competes directly with its licensees with its Pixel-brand phones complicates matters (see Chapter 21).

SAMSUNG

Samsung Electronics' smartphone product line-up covers the spectrum, from budget to premium. Continuing a theme introduced in Chapter 10, the key to understanding Samsung's position in smartphones is vertical integration, which it views as a source of competitive advantage. If an internal solution is not feasible, Samsung frequently calls upon local subcontractors it knows and trusts for specific parts. Companies that manufacture for Samsung often have no formal connection with the company but there may be historic links. This predisposition for making the components for the products it manufactures within its corporate orbit does have its downsides, as will become apparent. Impressive as the company is, Samsung cannot be best-in-class at everything.

Samsung Electronics is the largest member of the Samsung *chaebol*. Group companies are bound together by a complicated set of cross holdings. Samsung Electro-Mechanics (SEMCO) is technically merely an affiliate of Samsung Electronics. Listed on the Korean Stock Exchange, Samsung Electronics controls only 23.7 per cent of SEMCO's equity: the rest is owned by institutions and individuals. Yet, in operational terms, it is clearly a fully paid-up member of the Samsung Electronics group.

There can be no doubt about Samsung's broad-based technical prowess. In 2025 it spent $26bn on R&D. Because its capabilities are so wide and deep, Samsung sells components and parts to its smartphone competitors. Apple's reliance on Samsung for the majority of the display panels used in the iPhone is a well-known case in point. Samsung is the technological and market leader in OLED (organic LED) displays, which are taking over from LCD displays in smartphones. It also supplies memory chips to Apple. Samsung's overwhelming technical superiority in certain areas means that it can afford to ignore the generally accepted business rubric of not competing with your customers. If it were to adhere to this unwritten rule, Samsung would not be in smartphones. Because Samsung has so many fingers in so many pies, conflicts arise in its interactions with others, and not just externally. Appendix 4 contains examples of the difficulties that can arise in a complex environment such as smartphones where multi-product firms do business with each other.

Samsung Electronics has a complicated divisional structure. Smartphones fall under the MX (Mobile eXperience) division. On their own, phones probably accounted for 30 per cent of Samsung Electronics' revenues of $233bn in 2025. Since 2021, when Samsung adopted the current reporting structure, MX has recorded respectable pre-interest margins (9–12 per cent) from its mix of high- and low-end smartphone offerings and other mobile devices.

Financially, a far greater influence on Samsung Electronics than MX is the Device Solutions (DS) division. Device Solutions is essentially Samsung's semiconductor business. To reiterate the point made in Chapter 3, Samsung is a semiconductor juggernaut. As the leading integrated device manufacturer, it is a weighty competitor in every sector of the industry. Commodity-like memory chips account for two-thirds of the turnover generated by DS. Aside from memories, DS manufactures image sensors and the Exynos range of SoCs.

A crucial fact about Samsung – which is not always fully appreciated – is that the semiconductor business is the powerhouse that drives group profits. Samsung Electronics is like a car that is powered by an unfeasibly large engine connected to one wheel. Unfortunately, it is – especially in memory chips – an inherently cyclical business. The semiconductor industry generally is subject to cycles due to the enormous sums that need to be invested in additional wafer fab capacity ahead of demand. When the memory market is on an upswing there is much money to be made. The reverse is equally true.

The memory chip cycle turned down with a vengeance in the final quarter of 2022 and, in 2023, DS swung from a substantial profit to a massive loss. Fortunately, by the beginning of 2024, the upswing was on its way and DS made a decent profit for the year. The following table demonstrates the extent to which Samsung relies on its prowess in memory devices to drive profitability – in a good year.

OPERATING MARGIN	2019 margin	2020 margin	2021 margin	2022 margin	2023 margin	2024 margin	2025 margin
DS Division	22%	27%	31%	24%	Negative	13%	19%
Total group	12%	15%	19%	14%	3%	11%	13%

We referred earlier to Samsung's well-known policy of vertical integration. This 'everything under one roof' approach confers benefits, but it can also create tensions, both externally and internally. In the context of smartphones, the company's own Exynos SoC is a case in point. Samsung developed Exynos to challenge Qualcomm in the business of supplying SoCs to Android smartphone manufacturers. It would be natural to assume that the primary target for an internally designed SoC would be Samsung's own phone range. It is not clear to what extent Samsung's semiconductor division had this in mind when designing Exynos as opposed to meeting the needs of external handset manufacturers. Either way, it has not been a resounding success. There is a general view in the industry that Exynos has consistently failed to match the performance of Qualcomm's Snapdragon. The fact that the Samsung Galaxy range continues to use Qualcomm SoCs in large quantities rather than Exynos testifies to a lack of faith in its own product.

In smartphones, Samsung has carved out global leadership by marketing attractive, well-featured phones covering the entire price spectrum. Strangely, there is one country where it has failed to make an impact: China. Bluntly, Samsung 'lost' China. In 2014, the company was the uncontested No. 1 handset provider with an 18 per cent market share. That figure is now 2 per cent. How did this happen? There are several reasons, but chief among them are:

- After Google's 'expulsion' from China in 2010 Samsung made no attempt to build an alternative to Google Mobile Services acceptable to the Chinese consumer.
- Samsung executives were tin-eared in that they neglected to include features that the Chinese consumer liked. It was a failure to 'localise'.
- Retail outlets are important in China, especially for the sale of expensive products like the Galaxy S range. Samsung had virtually no physical presence.

Geopolitics also played a role. South Korea has been concerned in recent years not to become too dependent on China. Accordingly, Samsung has been winding down its production facilities there since 2018. From 63,000 employees in 2013, Samsung's Chinese workforce fell to 18,000 in 2022. The company's withdrawal did not go down well with the Chinese public and undoubtedly made the task of restoring its smartphone reputation harder.

As befits one of the largest companies in the electronics sector, Samsung has a structured and rigorous process of evaluation of potential suppliers. In view of the built-in preference for an internal – or at least Korean – solution, selling to Samsung is a tough proposition. Internal politics can play a role. The centre may want to bring on a new supplier, but may then encounter spirited resistance from the subsidiary making that self-same component.

APPLE

Apple could not represent more of a contrast with the multi-faceted leviathan that is Samsung. Phones account for just over half of Apple's revenue. The iPhone range is confined to the top end of the market in terms of both quality and price and is, of course, protected by Apple's own 'walled garden' software ecosystem. There is no denying that Apple's business model is highly successful – technically, commercially and financially.

Apple manufactures nothing internally. Procurement therefore is vital to Apple's operations. However, to describe Apple as an outsourcer is a misnomer. The company has completely redefined the concept to refer to a process that is the reverse of the conventional arms-length outsourcing relationship. Its involvement with its suppliers is all-encompassing, embracing all aspects of logistics and production. It is no coincidence that Tim Cook, Apple's CEO, made his name in this area. He joined Apple in 1998 as senior vice president for world operations, where he mastered the supply chain. For 11 years he worked closely with Steve Jobs, Apple's visionary founder. During Tim Cook's time as head of world operations, Apple kept costs down and produced devices more efficiently.

> 'From the start of his Apple tenure, Cook set colossally high expectations. He wanted the best price, the best delivery, the best yield, the best everything.' ('The job after Steve Jobs: Tim Cook and Apple', Yukari Iwatani Kane, *Wall Street Journal*, 28 February 2014)

Apple is not an out-and-out innovation pioneer. What it is adept at doing is letting other companies identify a market for a particular product and prove that demand exists. It then directs its formidable R&D and design resources towards coming up with an attractive and user-friendly device with broad appeal, which it sells to its existing customer base and beyond. The essential appeal of an Apple device lies in the software and the user interface (UI). The company has successfully replicated this formula several times, from the iPod to the iPhone and the iPad.

Financially, the Apple business model has many attractions. Apple engages in R&D and design but outsources component production and assembly. Some R&D costs are loaded onto suppliers while capital expenditure is normally transferred to the supply chain. Apple is supportive of its suppliers in terms of resources, such as training of employees, and sometimes financial assistance. There are certainly occasions when Apple will assist a supplier with funds to enable it to develop a component and set up production. If the supplier concerned is in the United States, the investment will come out of its Advanced Manufacturing Fund, which was established in 2017.

Generally, though, the company does not have to bear the heavy ongoing costs incurred by Samsung. As a consequence, Apple is asset light and highly profitable, with an operating profit margin that hovers around 30 per cent and a high return on capital. Apple's business model is, however, changing. Since 2017, far more resources have been devoted to R&D, such that its expenditure in 2025 ($35bn) exceeded Samsung's by a considerable margin. Similarly, investment in AI has led to substantially higher capital expenditure in 2025.

Apple is highly regarded for the quality and stringency of its processes among the global community of component suppliers. The path to becoming a qualified supplier is an extremely demanding one. It takes at least two years, a markedly longer timescale than that of its competitors. Apple is famous for meticulously verifying the entire company, including engineering design capabilities, quality control and manufacturing facilities. After qualification, the company is subject to a strict supervisory regime to ensure performance. Once qualification is achieved, however, the supplier is universally regarded as having met the highest standards. Appointment as one of the 200 suppliers Apple lists publicly each year (accounting for 98 per cent of all purchases) is an undoubted accolade and confers 'bragging rights' that can lead to additional non-Apple business.

Apple's overall thrust is towards greater ownership of both product and process. Consequently, the ability to design critical components to its own requirements is crucial. The next step is establishing the process whereby this custom component can be manufactured to its exacting specifications. (Rivals tend to just give the supplier a 'spec sheet'.) To achieve this, the company has vastly expanded its internal component design capabilities, which is reflected in the steep rise in R&D expenditure.

Each year, Apple publishes a list of its suppliers so that everyone knows who they are. An analysis by Bank of America relating to the fiscal year to 28 September 2023, disclosed that China remains its primary production base: 156 of Apple's suppliers had some manufacturing presence in China. Bear in mind this is by numbers. As explained in Chapter 7, by value, the Chinese contribution is considerably lower. An increasing proportion of components are manufactured elsewhere in South-east Asia, often by Chinese-owned suppliers. Apple has clearly signalled the direction of travel by encouraging greater diversity of supply, with the focus on India and Vietnam. Reducing Apple's dependence on China will, inevitably, be a gradual process, if only because so many lower- and mid-tech components in an iPhone can currently only be made at an acceptable cost and quality in China.

A key downside risk for Apple is that, for the company to succeed, the machine has to operate flawlessly. Each of its suppliers needs to play its full part. Apple does not control them and cannot be sure that they will not bow to competing demands. The upside is that it is a very large purchaser and therefore has leverage. For many smaller companies, Apple represents a significant part of their business. Apple is famously secretive about everything and insists on the same level of secrecy from its suppliers. An Apple Purchase Agreement is a fearsome document, with stringent penalties for any breach of confidentiality. Suppliers are also required to sign Non-Disclosure Agreements. Strenuous efforts are made to prevent details of new models leaking before they are announced. To no avail. The disparate and widespread web of suppliers makes this impossible and there are constant leaks about upcoming models.

Apple goes to absurd lengths to prevent suppliers mentioning it in public as a customer. Instead, they are forced to resort to coy phrases that fool no one. It is manifestly pointless because everyone knows it is Apple. On 4 March 2022 the *Wall Street Journal* – not a newspaper noted for its flippancy – could not resist exploring some of these. The heading read:

> 'The Big A? The Fruit Company? Why the Maker of iPhones Must Not Be Named.'

The article went on to say:

> 'It is the dominant American maker of smartphones, a household name to billions, and for many makers of high-tech parts their most important customer ever. Just don't ask who it is.'

Deep in TSMC's Consolidated Financial Statements for 2024, for example, we are told that 'Customer A' accounted for 22 per cent of net revenue in that year. How appropriate that 'Customer A' was Apple! (By 2025, Apple was down to 17 per cent of TSMC's sales, having been overtaken by Nvidea.)

As a matter of practice, Apple generally prefers every component in an iPhone to have at least two sources, for security of supply and to provide competitive tension to keep the price down. Generally, it seems that the template is a main supplier with 50 per cent of production, a second source with 30–40 per cent and then a third with 10–20 per cent. Suppliers often develop a component at Apple's behest for a new model. Apple knows what it wants, and the instructions are precise and exacting. There are inevitably times when a contractor is unable to perform, in which case the relationship is terminated. Apple is as commercially hard-nosed as any other in the electronics industry.

Apple is known for providing a consistent flow of business for its suppliers, as it constantly launches new and improved iPhones. The trade-off is that the margins it typically offers are not generous. Nor is the business guaranteed. This leaves a small, single product firm highly exposed. A large, multi-product firm can take the risk that the margin squeeze becomes too onerous or – as does happen – Apple decides to withdraw from the contract. Consequently, the vast majority of the firms in Apple's supply chain are multi-product operations.

There are several instances of Apple engaging in hardball tactics. During 2013 it contracted with a company called GT Advanced Technologies (GTAT) to make sapphire (rather than glass) screens for the next iPhone, loaning the firm $580mn to fund the startup costs. In the event, the iPhone stayed with glass and GTAT filed for bankruptcy the next year. In 2016, the British company Imagination, which supplied Graphics Processing Unit chips to Apple, was approached by Apple about a possible takeover. No offer emerged. Then, in April 2017, Apple announced it would stop using Imagination GPUs as it had developed its own. It also hired several Imagination employees. Such was the degree of surprise at this move, Imagination's shares collapsed by two-thirds on the day. The shares of Dialog Semiconductor, a British-German company, crashed in 2017 when it was revealed that Apple was intending to replace the IC it bought from the firm with one of its own. Doing business with Apple is very much a double-edged sword.

Apple, when it contracts with a company for a component, will try to insist on exclusivity – that this component will not be sold to any other smartphone manufacturer. This applies particularly if it incorporates a proprietary element that gives Apple an edge. Whether Apple is actually able to enforce exclusivity depends on the power balance between Apple and its supplier. In some instances, Apple has no choice but to treat the supplier as an equal. Apple has been designing its own SoC for the iPhone since 2013. Consequently, TSMC occupies a critical position as Apple's chipmaker. Sony is in a similar situation vis-à-vis Apple owing to its ability to make image sensors that others cannot provide. Sony has a strong relationship with Apple but that does not preclude it from selling to Oppo, Vivo, Xiaomi, Huawei and Samsung.

China has been an important source of growth and profitability for Apple, accounting for a fifth of group sales. The Chinese love affair with Apple products began in 2010. Consumers clamoured to buy iPhones and iPads. Ownership was considered to be a status symbol. Such was the demand in the ensuing years that Greater China (defined as including Hong Kong and Taiwan, no doubt in a nod to Chinese sensibilities) earned a materially higher margin than the Americas or Europe. The following figures extracted from the company's annual filing with the Securities and Exchange Commission for Apple's financial year to September 2023 illustrate the point.

Geographical area	Percentage of Apple's total net sales	Percentage of Apple's operating income before R&D expense	Operating margin percentage
Americas	42%	40%	37%
Europe	25%	24%	38%
Greater China	19%	20%	42%
Japan	6%	8%	49%
Rest of Asia Pacific	8%	8%	39%
Totals	**100%**	**100%**	**39%**

However, during 2024 this long run of above-average margins ended. Chinese enthusiasm for iPhones cooled, revenues fell 8 per cent in dollar terms and the margin on sales declined. Exactly why and how this happened is covered in Chapter 23.

It will be apparent from our discussion in Chapter 7 the degree to which Apple is dependent on China for production of the iPhone. It is not just that most assembly takes place there. It is the source of many critical components, frequently not high-tech in nature. Tim Cook, at the China Development Forum in Beijing in March 2023, described it in this way: 'Apple and China ... grew together and so this has been a symbiotic kind of relationship.'

How did Apple get to this point? History is important here. Thanks to a leak of internal Apple documents in December 2021 by the respected tech website, *The Information*, we now know much more about the background. In 2016 Apple was under regulatory pressure in China and product sales were faltering. Tim Cook made a series of visits there. Despite Apple's heavy investment in its supply chain, the Chinese authorities felt the company was not making enough of a contribution to the economy. To protect its market position, Tim Cook signed an agreement promising the company would do its part to develop China's economy and upgrade its technology through investments, business deals and worker training.

In keeping with Apple's penchant for secrecy, none of this was disclosed to shareholders. This is readily understandable from the company's point of view. President Trump's animus against China meant that it would have been politically toxic. For its part, Apple committed itself to more than $275bn in spending over five years. As part of the agreement, Apple vowed to place more of its business with Chinese companies, replacing others in Asia. The Apple supply chain indirectly supports more than 1.5mn jobs in China. In simple terms, the deal was: 'You invest in China and we give you more-or-less free rein in the domestic smartphone market.'

The 2016 agreement permitted Apple to sell iPhones in China with few restrictions, provided the company abided by local rules. In practice, Apple has had to fundamentally modify its operating procedures to meet Chinese demands. The iCloud in China is not controlled by Apple but by a Chinese company, AIPO Cloud (Guizhou) Technology Co., that could be ordered under Chinese law to provide the government with access to all of Apple China's customer data. The Chinese regulator also insisted on the removal of tens of thousands of 'offending' apps. None of this sits well with the company's heavily publicised commitment to user privacy. It is the harsh reality of doing business in China.

Apple has two objectives in China: first, to ensure the smooth operation of its assembly and component supply chain; and second, to preserve access to the Chinese smartphone market. To maintain its position in the face of superpower trade conflict, since 2018 it has engaged in an accomplished geopolitical high-wire act. What is curious is that, when the fact of Apple's 2016 overarching deal with China became widely known in 2022, the reaction from politicians and the media in the USA was surprisingly muted. Recently, though, a few Republican senators have become more outspoken. After all, through its actions, Apple has been – and still is – giving succour, both financial and technological, to a country that is now viewed as a strategic rival to the United States.

Most likely the reason for this willingness to overlook is that Apple is a de facto national champion. It is an icon, an exemplar internationally of American entrepreneurship and innovation. And, despite its extensive use of foreign labour, it is the source of considerable employment in the USA. According to the company's website, Apple is responsible for 2 million jobs: 80,000 direct employees, 450,000 through USA-based suppliers and 1.5 million developer jobs attributable to the App Store ecosystem.

Of even greater import, the company's current share price is several times what it was when Tim Cook did the deal with China in 2016. Seemingly, in America, financial success overrides all other considerations. Taking all this into account, it has to be assumed that Apple is 'given a pass'. A lawsuit brought by the US Department of Justice in March 2024 accusing Apple of monopolising the domestic smartphone market briefly dented the stock price. It will take several years to resolve. Most investors shrugged it off, believing it will have little long-term impact.

BBK ELECTRONICS

'BBK who?' will be the reaction of many readers. It may also come as a surprise that several of the leading Chinese brands are connected. BBK Electronics Corporation markets smartphones as Oppo, Vivo, OnePlus, Realme and iQOO. It also makes and markets television sets, audio equipment, vacuum cleaners, electric kettles, etc. Its headquarters are in Dongguan, a city just to the north of Shenzhen. Adding up the smartphones shipped by the constituent brands of BBK there is little doubt the company would overtake Xiaomi in worldwide sales.

Is it legitimate to treat BBK as a single company? BBK is not quoted, nor are any of the constituent companies formally or legally under its aegis (as far as one can tell). So there is minimal transparency and limited information. What is clear is that the BBK brands have a common origin. The constituent companies operate independently but share a common philosophy and it seems there is a degree of co-ordination. BBK's website – in translation – is strikingly uninformative, describing it as a consumer products company. There is only a glancing reference to smartphones.

BBK is the creation of one remarkable man, Duan Yongping. Now in his 60s, he has lived in the USA for 25 years. The story of why he resides in California is a delightfully romantic one. At the age of 38 he pursued the love of his life, Liu Xin, who had emigrated to the USA some years before and was working there as a photojournalist. She agreed to marry him on condition he moved permanently to the USA, which he did.

Born in China, Duan Yongping obtained a degree in electronic engineering and then worked in a state-owned vacuum tube factory. In 1990, Duan joined a company where he proved to be a highly effective manager. When, in 1995, he found himself in conflict with management on corporate strategy, he left with six fellow employees to form his own company, BBK Electronics. BBK soon became a significant manufacturer of DVD players and of dumbphones. When Duan Yongping moved to the USA, he left four disciples behind to run the business. He has continued to communicate with them from afar.

Fierce competition in dumbphones in 2005 almost bankrupted the company – and brought Duan back from California for a time. To dissociate itself from the past, BBK established its Oppo brand at this time, led by one of Duan's protégés. Oppo initially made and sold music players and then moved into smartphones. Vivo was founded in 2009 by Duan and another of BBK's executives. It was followed in 2013 by OnePlus, a spin-off from Oppo. In 2018 a vice-president of Oppo, and head of India operations, set up Realme. Then, in 2019, Vivo spawned a further sub-brand, iQOO. The image below shows Duan Yongping with his creation.

Source: 'BBK Electronics: A Leading Innovator in the Tech World', Asma urooj, Medium, 16 January 2022

The people who founded these companies may, or may not, have worked directly with Duan Yongping but all had been imbued with his way of doing business. They operate independently but follow a common set of principles and share resources. Two examples: OnePlus has leased space on Oppo's assembly line; and Realme benefited from Oppo's manufacturing capability and retail distribution network when setting up in India. BBK, in keeping with Duan's manufacturing background, assembles most of its smartphones internally.

BBK companies seek to develop attractive handsets with solid, well-accepted features at a series of price points below Samsung and Apple. Innovation is essentially low risk and incremental. A core tenet of Duan Yongping's philosophy is the belief that business partners – employees, suppliers, distributors, retailers – deserve to be treated properly and fairly. Opacity makes it impossible to establish the precise nature of the financial relationships within BBK but, reportedly, another element in Duan's modus operandi is that the constituent companies are majority owned by their employees. Another common approach is high profile, ubiquitous promotion to differentiate their phones from the pack, often involving celebrity endorsement, product placement, sponsorship and social media. The following quote captures the flavour of BBK's saturation marketing approach.

> 'Vivo and Oppo, in particular, have developed the 100 million strong south-east Asian market by tapping star power. Last May, Vivo struck a deal to become the official smartphone sponsor for the World Cup in 2018 and 2022 … Meanwhile, Oppo advertisements have virtually monopolised entryways around Bangkok's central business district, as well as the Sukhumvit underground rail station. At Jakarta's ITC Kuningan shopping mall, the section dedicated to mobile phones is blanketed with green Oppo signs, along with blue Vivo signs.' ('Chinese smartphone makers in all-out blitz for Southeast Asia', *Nikkei Asia*, 16 May 2018)

The distinct marketing profiles of the BBK brands are set out below. Inevitably, as time goes by, it has become more difficult to avoid overlaps and the cannibalisation of one brand by another. The BBK brands also differentiate themselves by skin. For example, Oppo has ColorOS, Vivo offers OriginOS and OnePlus OxygenOS.

» **Oppo**. As the first BBK smartphone contender, Oppo is focused on China and the rest of Asia, and used celebrities and product placement in movies and TV series to become known.
» **Vivo**. The first Vivo phones appeared in 2011. The focus was on ultra-slim, feature-packed, mid-range devices with an emphasis on innovative features other than the camera.
» **OnePlus**. Set up in 2013, OnePlus pursued an alternative approach, targeting both Indian and Western markets via the internet and social media. OnePlus offered tech-savvy consumers high-spec phones at a competitive price. In its marketing, it projected the image of a rebellious upstart. In 2020 it moved into the mid-price market segment with its Nord range.
» **Realme**. From a standing start in 2018, Realme has carved out a leading position in Asia's emerging economies with entry-level but high-spec smartphones targeted at young techie buyers via social media. Realme's principal market is India, but it sells well in all other Asian markets.
» **iQOO**. Vivo felt it too needed a non-mainstream sub-brand like Realme and, in 2019, created iQOO, aimed at a younger demographic, interested in gaming and streaming, which buys online.

Publicly, BBK companies deny other than the most basic connection, but it is difficult to resist the conclusion that some kind of guiding hand is at work here. The suspicion must be that, at BBK's headquarters, there exists a cadre of Duan disciples monitoring and adjusting the structure of the BBK empire as they deem necessary to maintain its performance. Evidence for this comes from the announcement in 2021 that OnePlus would be reabsorbed into Oppo as a sub-brand after eight years of independence. OnePlus was increasingly encroaching on Oppo's territory and the duplication of resources made less and less sense. However, after a backlash from users, it was announced that Oppo and OnePlus will retain their separate Android skins.

As for Duan himself, he lives quietly with his family in Palo Alto, California. To what extent he advises BBK is unknown. He does apparently attend some board meetings. Duan has long been an admirer of Apple and, in January 2011, made an initial investment. Apple's stock price has increased around tenfold since then. He and Tim Cook of Apple have met several times. Apple and his other stock market investments have made him extremely wealthy, with a net worth of several billion dollars. Duan Yongping now thinks of himself, and is known throughout China, as a successful stock market investor and philanthropist. Though reclusive, he is a blogger on investment matters with 400,000 followers. Some call him the Warren Buffett of China.

XIAOMI

What is apparent from even a casual glance at Xiaomi is how dramatically different it is from BBK. Xiaomi is transparent and is willing to talk about its operating processes. There is much written about it online. It has been quoted on the Hong Kong Stock Exchange since 2018 and the information given to investors approaches that provided by Western companies. Xiaomi's 2025 annual report runs to 415 pages. Like BBK, Xiaomi is a Chinese corporate success story though the route that it took to get there is a more conventional one. Founded in 2010, and headquartered in Beijing, in December 2010 Xiaomi raised $41mn in venture capital from institutional investors and corporates, including Qualcomm. Xiaomi released its first smartphone in 2011. Xiaomi went public – while still loss-making – in 2018 at a valuation of $54bn. The share price fell below the IPO price immediately on listing, only exceeding it in July 2020.

Xiaomi was set up by Lei Jun with six other self-confessed computer geeks. A feature that distinguishes Xiaomi from BBK is the background of its original cadre of executives. Of the other five co-founders, four had studied in the USA and three had worked for Western companies. As such, they were familiar with Western norms and practices. Graduating in computer science in 1992, Lei joined Kingsoft, a Chinese software firm, and rose rapidly to become CEO. After Kingsoft's IPO on the Hong Kong Stock Exchange in 2007 he left to become an angel investor. Lei is a serial entrepreneur, having launched several companies in his career, and is a board member of several technology businesses that appeal to him. He models himself on Steve Jobs – right down to his studiedly casual attire, as below.

Source: 'Xiaomi CEO Lei Jun Makes $30 Million in Sales in Livestreaming Debut', Ding Yi, Caixin Global, 18 August 2020

Xiaomi rapidly gained market share to become the largest smartphone company in China by 2014. Having secured its home base, it swiftly expanded into Asian and international markets, spearheaded by an executive recruited from Google. Nearly half of Xiaomi's revenue comes from outside China, including a significant presence in European markets.

Having started in low- to mid-level handsets, Xiaomi is now placing more emphasis on the more profitable premium segment. The year 2022 was a tough one for Xiaomi, as indeed it was for all Chinese manufacturers. Handset sales fell sharply. Both sales and profits declined. In common with its local competitors, Xiaomi countered these trends with a premiumisation strategy. The following year and 2024 witnessed a solid recovery.

It is instructive to contrast the background of the two entrepreneurs. Duan came to mobiles from hardware while Lei's prior experience was firmly rooted in software. In fact, the software interface it uses on its smartphones was launched before its first phone. In August 2010 Xiaomi introduced an Android skin called MIUI in the form of firmware, which is embedded in the memory of every Xiaomi smartphone. Most Chinese manufacturers at that time simply added the Chinese version of stock Android with minimal customisation. MIUI has been instrumental in generating great customer loyalty in China through the Mi Forum, an online community of several million registered users. Engagement is encouraged. MIUI is highly customisable and allows thousands of keen users to invent new features every week. Xiaomi enjoys a cult-like status among its fan base. Lei Jun personally plays a crucial role here and elsewhere on social media, with 21 million followers on Weibo, the Chinese equivalent of X.

Lei Jun's desire to model himself on Steve Jobs went beyond simple hero worship. In the beginning, MIUI owed much aesthetically to iOS and similarities remain. The design of Xiaomi's handsets has been described as minimalist chic, which again draws comparisons with Apple. In 2025, Xiaomi went as far as renaming its phones so that the numbering system aligned with that used by Apple.

In 2025 41 per cent of Xiaomi's turnover was generated by smartphones. These are sold under the Xiaomi, Redmi and POCO labels. Redmi, launched in 2013, is a budget phone brand while POCO is aimed at gamers and does well in India. Initially, Xiaomi made extensive use of the internet to market its phones. It has since added thousands of retail customer centres to the selling network but only in high-footfall locations, like malls.

From the beginning, Lei Jun had ambitions to turn Xiaomi into a broad-based consumer products company and the percentage of revenue attributable to the sale of handsets is steadily declining. Beyond smartphones, Xiaomi's business comes from a wide range of smart products and services directed, in the first instance, at its hardware fan base. These include laptops, TVs, refrigerators, air conditioners, wearables and lifestyle products, such as electric toothbrushes and e-scooters. Being able to offer a variety of consumer products in addition to phones has made Xiaomi's retail outlets more viable. In early 2024 Xiaomi launched an electric vehicle (EV), which had been under development for three years. The company describes its corporate strategy as 'Human X Car X Home'. The EV was particularly well received by the Chinese public and investors responded accordingly, propelling the share price to new heights, well above the IPO price.

Reflecting Lei's background as an active private investor, Xiaomi has always taken stakes in a wide range of small, often startup, companies of interest to the group. The company's investment portfolio numbers several hundred firms. Part of the reason for the proliferation of minority stakes is the desire to bring these startups into the Xiaomi ecosystem. The company provides affiliated companies with a route to market via its own brand and distribution system.

Many of the products sold in Xiaomi retail outlets are designed to be connected to other group products. This diagram below illustrates the scale and breadth of the company's ambition as expressed at the time of the company's IPO.

Source: 'Xiaomi Today' website, 22 November 2018

Xiaomi is at the forefront of the Internet of Things (IoT) concept, which envisages the interconnection of smart appliances in the home or on the move. Typical examples are internet-enabled products, such as refrigerators, robot vacuum cleaners, home security devices, air purifiers and rice cookers, allowing users to control these from their phone via an app. In pursuit of its IoT strategy, Xiaomi announced a major step forward in October 2023: the replacement of MIUI with HyperOS. Seven years in development, HyperOS is more of an operating system than a customisation layer. It represents a fusion of Android with Xiaomi's proprietary IoT platform, Vela. HyperOS remains compatible with Android and links all the company's devices, whether at work, at home or in the car.

Since 2019, as the split between China and the USA has widened, Xiaomi has directed more of its corporate venture capital investment towards securing its supply chain. Xiaomi, unlike Huawei, has not been the target of US sanctions. Rather the move is perceived as an insurance policy in case the situation changes to its disadvantage. In March 2021 it was reported that Xiaomi had stakes in at least 30 chip developers and equipment firms and 25 other electronics hardware businesses. Then, in 2023, Xiaomi set up a $1.4bn partnership specifically to offset the impact of the US restrictions, which has invested in companies active in AI, smart manufacturing and automotive technology.

HUAWEI

Huawei is not the global force in mobile devices that it was in the middle of 2020. Until then, the company's smartphone shipments had been on a strongly rising trend. In Chapter 7 we observed that Huawei has, since May 2019, been the target of American sanctions, directed at its telecommunications arm but with a spillover effect on its smartphone business. Huawei's smartphones were subjected to a double whammy, hitting both software and hardware. First, the American administration instructed Google to deny the company access to the standard Android pre-installed apps, such as Gmail, Maps and YouTube, and to the Google Play Store. This has severely impacted handset sales everywhere with the exception of China, which has its own app ecosystem. Second, the inability to access key chips, specifically its own-design SoC, has seriously affected phone production. The events that led up to these restrictions are covered in Chapters 18 and 22.

Despite the impact of sanctions, Huawei has remained profitable, albeit margins have been squeezed. Huawei did not spend time bemoaning the embargo but resolved to find ways round it. This is an objective it eventually achieved – exactly how (with implications for CML) is the subject of Chapter 26. A terse statement from Huawei in August 2021, when sanctions began to have a real impact, referred merely to 'challenging times'. The Consumer Business Group took the biggest hit. Consumer (broadly synonymous with smartphones) generated 50 per cent of group sales in 2020. In 2025 it accounted for 39 per cent

In spite of these pressures, Huawei remains a formidable and energetic worldwide competitor. The company reportedly has a ferocious work ethic (the 996 schedule, referring to the Chinese tech sector's regime of 9am to 9pm, 6 days a week). Within China, the company is widely recognised as a national tech champion. Huawei has moved beyond its original strongholds in telecommunications – in which it ranks No. 1 worldwide – and smartphones into cloud computing, artificial intelligence, digital power, infrastructure and what it calls Intelligent Automotive Solutions. This latter unit does not make cars but sells its digital automotive technology to EV manufacturers. What these new activities have in common is (with the exception of AI) that they are less reliant on the advanced, sub-10nm mode chips that are the subject of the US restrictions.

In research and development, Huawei is undoubtedly a force to be reckoned with. Its commitment to innovation has tripled over the last ten years, from $9bn in 2015 to $28bn in 2025 (representing 22 per cent of sales). It is instructive to view Huawei's rapidly rising R&D expenditure in a global context. The table below shows that the company is now the sixth largest investor in R&D worldwide, ahead of Samsung though still well behind the American tech behemoths, buoyed up by the rush into AI.

COMPANY	2025 R&D EXPENDITURE $BN
Amazon	108
Alphabet (Google)	61
Meta	57
Apple	35
Microsoft	32
Huawei	28
Samsung	26

Source: Company annual reports, "The 2025 EU Industrial R&D Investment Scoreboard", European Commission, 2025

Huawei is, indisputably, a technological powerhouse, holding 165,000 patents worldwide. As a result of its strong patent position in telecommunications, it is a major licensor and, alongside Ericsson, Nokia and Qualcomm, has been responsible for creating global standards for 5G. Apple and Samsung are obliged to pay royalties to Huawei for each phone sold. Huawei earns more in patent fees annually than it pays out to other companies.

There can be no doubting Huawei's desire to be viewed as a premier international tech corporation, on a par with the likes of Google, Microsoft and Meta. In 2025, the company opened a brand-new $1.5bn R&D centre outside Shenzhen. When it reaches full capacity, 25,000 people will work there. Spanning 300 acres, with numerous cafes and restaurants, the lakeside campus is designed to foster an atmosphere of creativity and collaboration,

The founder of Huawei, Ren Zhengfei, who is now over 80, spent almost 20 years in engineering roles within the People's Liberation Army, giving rise to the suspicion that Huawei is subject to military influence or even direction. Huawei's hard-driving culture owes much to Ren's long stint in the military. He joined the Communist Party rather late, having been denied membership earlier because his father was a 'capitalist roader'. Ren founded Huawei Technologies in Shenzhen in 1987 and, by 1992, had developed and was selling 4G telecommunications network equipment. It is widely believed that his links with the Chinese Army and the Communist Party helped him to obtain government contracts, particularly in developing nations where Huawei has achieved a high level of penetration. Equally, it is undeniable that Huawei offers quality equipment at a competitive price.

Huawei is not listed on a stock exchange but – at first glance – provides information of a standard akin to that of a quoted Western multi-national with over 200,000 employees. Annual reports in English, containing a review and accounts audited by KPMG, have been published since 2006. The corporate website is comprehensive to the point of wordiness, as is the 189-page 2025 annual report. Both are long on generalities and short on numbers. Huawei sees itself as a broad-based technology enterprise with a mission to help industries 'go digital, intelligent and green'. The annual report contains much about Huawei's capabilities and technological ambitions but very little about the business and its performance – not something that a listed company could deny its readers. This makes perfect sense in the context of its target audience. The report is not directed at shareholders but suppliers, customers, collaborators, officialdom and the Chinese public at large. The object of the exercise is to offer these groups re-assurance, not to highlight difficulties. The American embargo is barely mentioned and then only in the most oblique manner. In his 2022 review the rotating chairman said:

> 'Plum blossoms tend to grow sweeter from a harsh winter's freeze. Today, Huawei is like a plum blossom.'

The other key problem from the point of view of an outsider is that there is no real information on Huawei's ownership. The 2024 annual report merely says:

> 'The group's ultimate controlling party is the Union.'

Shareholding employees are members of the Union and are entitled to vote. There is a complex system of corporate governance. Ren Zhengfei is still technically CEO though formally only one of the directors. Deep in the 2025 report we are informed that Ren Zhengfei owns 0.59 per cent of the company's total share capital. Aside from the Union, he is the only owner mentioned. Ren has consistently set his face against an IPO, believing that the company would be subjected to short-term pressures and that talented employees would cash out and leave. Arguably, if Huawei were listed, an open share register would help to allay some of the fears about where true control lies.

The Consumer Business Group sells a range of electronic products, such as PCs, laptops and smart speakers, and is pushing into IoT but smartphones account for the bulk of turnover. Huawei introduced its first smartphone in 2009, powered by its own Android skin called EMUI. The flagship P and Mate series phones have been well received globally, with plaudits especially for the excellence of the camera. Targeting the mid-market, Huawei introduced the Honor range of online smartphones in 2013, followed by the Nova range aimed at younger consumers.

In 2016 Huawei announced its intention of becoming the world's largest smartphone manufacturer. The company's ability through its fabless semiconductor subsidiary, HiSilicon, to design its own Kirin SoC for the P and Mate series was considered to be a source of competitive advantage in its bid for smartphone primacy. The Kirin SoC was fabricated by TSMC in Taiwan. In the following four years, the company's handsets made substantial inroads all around the world with the exception of the USA, largely a no-go area for Chinese handsets (for reasons discussed in Chapter 6). Huawei fleetingly achieved its ambition to be global No.1 smartphone vendor in 2020, shipping more phones than Samsung in the second quarter of the year.

Huawei's biggest problem since the imposition by the American administration of restrictions on TSMC has been its inability to manufacture its own-design SoC. This was a devastating blow, cutting Huawei off from its fabrication partner. China is actually the world's fourth largest producer of chips, ahead of the USA, but its problem is that domestic foundries are generally only capable of processing ICs of nodes of 28nm and above. As we have noted previously, flagship smartphones demand the smallest possible nodes, down to 5nm or below, for maximum performance and efficiency. Fortunately for Huawei, in November 2020, the American administration relaxed the chip restrictions it had previously imposed by allowing Qualcomm to supply its SoC for Huawei handsets, though not the latest versions supporting 5G capability. This represented some minimal relief. Lacking 5G, Huawei phones remained at a serious disadvantage in the marketplace.

How did Huawei respond to these sanctions? It first took action on the software front: the replacement of EMUI with an Android-based operating system not subject to the restrictions imposed by Google at the behest of the US government. Announced in 2019, HarmonyOS is billed as the world's first operating system dedicated to the era of IoT. It is claimed to function seamlessly across a wide array of devices and household items. HarmonyOS is best characterised as an ultra-thick skin. It is essentially what the industry calls a 'forked version' of Android, not a replacement. Using the same open-source code base as Android, it has been possible to run Android apps on the Huawei OS. Huawei offers its own version of the Google Play Store which it calls the AppGallery.

In 2024 Huawei introduced a revised version of HarmonyOS called HarmonyOS NEXT that is not compatible with Android. HarmonyOS NEXT (subsequently renamed HarmonyOS 5 and then 6) is designed to wean Huawei users off Android entirely. Huawei is working hard to persuade developers to write apps for this new version. By late 2025, according to the company, HarmonyOS had 27mn users, 10mn dedicated developers, and more than 30,000 apps were available on the HarmonyOS platform. Domestically, HarmonyOS is making progress as the major Chinese internet companies build new versions of their apps, though not without difficulty as Huawei's self-developed OS is fundamentally different from Android. However, without easy access to the Google Play Store, the vast majority of international consumers will be disinclined to make the effort to download apps from elsewhere. Expanding internationally with an alternative set of apps on an alternative operating system – however good – is an uphill task.

17.

Samsung and Hutch

FORWARD WITH SAMSUNG

At the end of Chapter 13 we left CML early in 2015 confronting the disappointment of Hutchinson but deep in conversation with Samsung, exploring the possibility of a mutually beneficial collaboration that would see SMA OIS adopted across the Samsung range. From CML's point of view, getting into bed with Samsung Mobile – with its high volume and its structured approach to suppliers – was preferable to having to deal with entrepreneurial, free-wheeling Chinese manufacturers (Huawei excepted). When Samsung finally signed a development licence with CML, the roadmap envisaged a slot for SMA OIS in the next generation of Samsung's premium Galaxy range.

There was, inevitably, a fly in the ointment. Samsung was not happy with the prospect of relying on Hutchinson Technology Inc. for the actuator. Samsung lacked faith in HTI's ability to perform. True to form, there was a preference for a manufacturing source from within the Samsung network. Quite apart from the information provided by CML, HTI was a public company. Its stuttering attempts to get into production, and its financial weakness, were plain for all to see. The original agreement between CML and HTI did not permit a third party to access SMA. But the agreement had been changed. Under the new terms, CML could permit Samsung to access its technology directly.

Nearly two years after the start of discussions, in December 2015, Samsung signed a production licence agreement. Much time had been spent resolving the tangled tripartite situation between CML, HTI and Samsung. In the event, HTI was cut out of the equation. The actuator would be produced entirely within Samsung's existing supply chain, an arrangement that included its affiliate, SEMCO, and a Korean contractor, Sungwoo Electronics. The actuator would be manufactured by Sungwoo and then sent to SEMCO for assembly into a camera module. CML received $1mn as an upfront payment, 50 per cent of which was an advance against royalties. The mood was upbeat, such that there was even talk of Samsung taking an equity stake in CML.

Everything was focused on inserting CML's actuator technology into Samsung's next flagship phone, the Galaxy S7. Before committing to mass production, Samsung insisted on a trial run, to see if the actuator could be produced at the rate and at the cost required. The S7 was announced in February 2016. However, the way in which OIS was presented by Samsung to the public contained a peculiarity – a strange reference to 'Smart OIS'. The reference mystified the techno-commentariat, as this quote from a well-known website demonstrates:

> 'The only thing we're not sure about is what Smart OIS stands for. There's no specific mention of **optical** image stabilization on the Galaxy S7 and S7 Edge and the less protruding lens could mean it's only Electronic image stabilization (EIS). We'll certainly have to see about that before we pass our final judgement.' ('Here's an in-depth look at the Galaxy S7 camera', GSMArena, 21 February 2016)

In March 2016 CML was optimistic that production of SMA actuators by Sungwoo would start in April and be phased in to replace the existing VCM actuator by July. A major concern was yield. Even with substantial on-the-ground assistance from CML engineers, there was no guarantee that Sungwoo could achieve the necessary yield level in a compressed timeframe. (HTI was finally achieving yields of 85 per cent or more but it had taken considerable time to get there.) And yield, of course, went straight to cost.

Samsung was not prepared to qualify SMA for use in the S7 if the cost exceeded that of the incumbent VCM actuator. Disappointingly, management continued to delay sign-off and the confirmation date slipped into the summer and then into the autumn. Technical problems unconnected with CML also slowed things down. Yields continued to improve but were still short of what Samsung wanted. There was, however, a bright spot: Samsung confirmed that SMA OIS had been qualified for inclusion in the soon-to-be-launched Galaxy Note 7.

FIRE DOWN BELOW

Then, in August 2016, an incident occurred that threw the whole of Samsung Mobile into turmoil – with profound consequences for CML. The Note 7, a larger version of the S7, was released in mid-August. The new model was well received and sales climbed. But, over the next three weeks, there were disturbing reports of Note 7 batteries catching fire. Videos of 'self-combusting phones' went viral. Then, on 2 September, Samsung announced a global recall of several million devices, placed a software block on devices in the field and directed all available resources towards identifying the problem and finding a solution. It was as if fire had broken out in the hold of a ship and every available member of the crew had been sent below to put it out.

Seven hundred engineers were diverted from other tasks to conduct internal testing of handsets in a specially built facility. Samsung suffered serious reputational damage while all this was going on, which threatened to have an impact on the entire range. Then, in January 2017, after four long months of silence, the company announced it had found the cause and generated a fix. The problem was identified after testing 200,000 phones and 30,000 batteries. It turned out that there were errors in both the design and the manufacture of batteries supplied by two different entities, one of which was a Samsung affiliate.

Subsequently, there was a managerial shake-up in Samsung Mobile and heads rolled. CML had several people on the ground in Korea, working with Samsung and Sungwoo. Personal relationships were inevitably disrupted. Stung by this all-too public failure, Samsung insisted on more stringent testing of *every* component. The entire organisation became markedly more risk averse. Consequently, the Note 7 battery problem proved to be a major setback for CML. Nevertheless, discussions continued with Samsung on a product roadmap for SMA OIS. Samsung indicated they wished to see CML's actuator in the forthcoming Galaxy S8, but the timing was uncertain.

What happened next is unclear, but the upshot was that SMA OIS was not designed into the S8. 'Smart OIS' appeared – again without explanation – on the front of the S9 but was quietly dropped in favour of plain 'OIS' for the S10. The CML–Samsung relationship never recovered from these initial setbacks and simply faded away. Why did it fail? Most of the blame can be attributed to the singularly unfortunate circumstance of the battery problem. Sungwoo and SEMCO had proved they could produce the camera module containing the actuator. Then, just as it was preparing to qualify a new type of lens-moving mechanism, Samsung was forced into crisis mode for several months. Timing also played a role. Early in the discussions, SEMCO expressed interest in a manufacturing licence for SMA, when the terms of the agreement with HTI prevented CML from doing so. By the time the agreement was changed, SEMCO had moved on to invest further in VCM. At root, it appears that the decision not to go ahead with SMA reflected a recalibration of risk by Samsung Mobile in the second half of 2016 and beyond.

Ironically, Samsung may also have been influenced by CML's promotion of a novel concept: a combined AF and OIS solution using eight wires instead of four ('8-wire'). (We noted the genesis of this idea in Chapter 11.) Putting the two together promised a simpler design and substantially lower cost. For Samsung, it opened up the exciting prospect of extending OIS beyond premium phones to mid-market handsets. However, as yet, the 8-wire actuator only existed on the drawing board and, in any event, it would require a completely new licence arrangement. CML, while pushing to get into Samsung, had not neglected opportunities elsewhere. Chinese companies were rising stars in the smartphone firmament and CML had been focusing on opportunities there while grappling with Samsung. Huawei soon emerged as the front runner.

THE END OF HUTCH

Hutchinson Technology Inc.'s demise as an independent entity was neither clean nor quick. The acquisition by TDK was scheduled to complete in the first quarter of 2016. HTI's survival was at stake, as it was committed to repaying the great bulk of its loans by January 2017, with no prospect of being able to refinance. Beyond that, bankruptcy loomed. Unfortunately for HTI, the US anti-trust authorities objected to the degree of consolidation that the 'merger' would bring to the Hard Disk Drive (HDD) industry. Negotiations dragged on and, for a period, it was touch-and-go whether the deal would be done in time. In the end, it was completed on 5 October 2016.

Why did TDK want HTI? HTI's position in SMA OIS was a secondary consideration to the rationalisation of global HDD capacity but undoubtedly figured in the calculation. TDK's VCM actuator business was struggling, and the company was already working with CML and HTI on SMA. The prospect of utilising the excess capacity in suspension assemblies to manufacture SMA actuators had definite appeal. Overall, the acquisition had all the appearances of a good fit with TDK's existing business.

18.

SMA spreads its wings, hope from Huawei, geopolitical undercurrents

HANDSETS AND BEYOND

The advent of multi-camera phones in 2017 was a distinct positive for CML. Each camera module offered a different capability. Some would require actuators for stabilisation. Zoom lenses – which amplify shake and therefore fell right within CML's field of interest – were on the horizon, as were wide-angle lenses requiring larger sensors. Smartphones remained central for CML but the team began to look more broadly at opportunities beyond handsets. The company had always viewed SMA as a technology platform, a base on which it could build a business providing solutions for a diverse set of needs.

Several such opportunities to embed SMA actuators in products other than smartphones emerged after 2015. Generally, customers approached CML.

» **Drones**. Cameras on drones have to endure extreme vibration. Actuator Solutions GmbH (ASG), the joint venture between SAES Getters and Alfmeier set up in 2011, shipped, under licence, a substantial quantity of SMA AF actuators to DJI, the Chinese company, that then dominated the drone market, and still does.

- **Special application phones**. A manufacturer of ultra-reliable, ultra-rugged smartphones for use in hazardous environments, Sonim Technologies, contracted with CML for the supply of SMA OIS actuators for the camera in a new model.
- **Wearable devices**. Ubiquiti Labs, the subsidiary of a maker of Wi-Fi products, decided to enter the wearables market with an inconspicuous pendant camera called FrontRow. It was aimed at those wanting to capture instant photos or video. SMA OIS was designed in.
- **New 'form factor' phones**. Undeterred by the Nokia debacle, Microsoft decided to re-enter the handset market with a revolutionary concept, a two-screen, hinged phone. Sensibly, this time the company opted for Android as the OS rather than Windows. This form factor (as any phone design concept is referred to in the trade) required a thin camera and Microsoft approached CML to discuss using SMA for its AF and OIS functions.
- **Pneumatic valve for lumbar support**. Actuator Solutions GmbH started selling an SMA actuator to control the lumbar support mechanism in car seats, generating useful, if modest, royalty revenue.
- **Haptics**. The science of touch was a logical diversification, as it had been used in smartphones since 2011. Haptics are any type of technology that gives you a tactile response – for example, when your phone vibrates. The word has its origins in the Greek words *haptikos* (able to touch) and *haptesthai* (able to lay hold of). Haptics allows non-responsive surfaces like touchscreens to simulate the feeling of using real objects like buttons and dials. This can involve vibrations, motors and even ultrasound

beams. CML engineers designed an SMA haptic 'button' that was smaller than current solutions and showed a prototype to prospective customers.

» **3D sensing.** A further promising area, prompted by the arrival of multi-camera phones, was three-dimensional (3D) sensing. This is the process of capturing a real-world object's length, width and height with clarity and in-depth detail. It augments the camera's capabilities for facial and object recognition. More exciting than 3D sensing itself was its potential as a route into the seemingly boundless world of augmented reality. AR – as it is invariably called – combines 3D images with computer-generated projections to supplement what the eye sees. It superimposes virtual objects on the physical world as, for example, when a piece of furniture being considered for purchase is 'placed' in position on a (handset) screen with a view of the room you are standing in. The team at CML had high hopes for 3D sensing, believing that SMA had a role to play.

» **Medical devices.** During 2017 CML was awarded grants to work with two leading British universities to explore the possibility of using SMA in medical devices. Specifically, the projects focused on actuation in surgical tools.

» **Depth mapping technology**. CML was approached by Intel to engage in a project on depth mapping.

Truly, SMA was spreading its wings, offering CML a cornucopia of possibilities.

HUAWEI RISES

China continued to be a hotbed of smartphone activity. In parallel with the drive to capture Samsung as a customer, CML continued to pursue this expanding market via the actuator manufacturing licensing relationships it had established with New Shicoh and TDK. Both were intent on marrying their VCM autofocus actuators with CML's SMA OIS actuator so as to provide a package for camera module makers.

At the end of 2015, HTI was shipping sample quantities of its new and improved Gemini actuator, hoping they would pass the reliability tests demanded by the camera module integrators. There were indications of interest in Gemini from smartphone customers, notably Huawei, Oppo and HTC, but nothing firm. Huawei was looking for options to replace the unreliable VCM OIS actuator it was buying from TDK. For its part, TDK was anxious to use SMA to restore its reputation with Huawei and was pressing hard for it to be designed into the next Huawei flagship phone, the P9, due out later in 2016.

In the event, time was too short for CML to obtain a design win for the P9. Huawei instead encouraged the CML/TDK partnership to aim, in the first instance, at something more modest. Dual cameras were the latest hardware innovation. Conscious that two cameras on one phone created new risks, and SMA was an untried technology, Huawei was keen to try out SMA OIS as a second source to VCM on a low-volume existing handset first. It was also concerned about reliability of supply if it committed to SMA mass production. Huawei therefore pressed for at least two sources of SMA actuators to be in place when it got to this stage.

It is worth recalling where Huawei stood competitively in early 2016. Its star was rising. Huawei handsets had acquired a reputation for quality along with advanced photographic capabilities. The company sold 140mn handsets in 2016, putting it in third place globally. Nearly half of its smartphone sales were outside China. And it was planning to make and sell 200mn handsets in 2018. Market penetration had improved everywhere, including Europe, though, for the reasons given in Chapter 6, not the USA.

At this point the camera actuator market was in ferment. There were doubts about VCM's ability to perform in multi-camera phones. One concerning issue was the possibility of magnetic interference with other components. (SMA, of course, contains no magnets.) Sunny Optical's reaction was typical. As a camera module integrator, Sunny's preferred actuator supplier was a Chinese company called Billu (full name: Shanghai B.L Electronics Co. Ltd). Billu was a significant force in the Chinese VCM actuator business. Worried they might miss the boat on SMA, Sunny asked Billu to work with CML. The result was a licence agreement, similar to the one that had been signed with TDK.

After a while, relationships between the different parts of the camera module supply chain settled down. Every camera market participant could see the implications of two – and more – modules per phone for the future of the industry. Of the Chinese smartphone vendors that CML had in its sights, Huawei turned out to be the most responsive. It had ambitious plans for smartphone photography. The prospect of aligning its leading-edge photographic capabilities with CML's innovative OIS actuator technology had distinct appeal.

At the top end of its range, Huawei had settled on a 'two flagships a year' timetable. The P series appeared in the spring followed by the Mate series in the autumn. Generally, the difference between the two Huawei handsets is that the P emphasises photographic performance and design while the Mate has the newest processor and display.

HITCHED TO HUAWEI

In 2018, CML finally achieved its objective. After more than two years evaluating SMA, Huawei decided that CML's actuator had more to offer than VCM in the context of the company's longer-term vision of what they believed could be accomplished in smartphone photography. In short, it offered a path to a superior actuator. CML's efforts were rewarded with a firm contract for the P30 Pro, the higher end model in the next iteration in the P series, due for launch in early 2019. Indeed, the relationship with Huawei was now such that it told CML it intended to include SMA OIS in the primary camera of the next Mate Pro due later in the same year.

To secure a slot on a high-end model of a leader in the smartphone business was a breakthrough of major proportions. For the first time, SMA OIS was an integral component in a handset that sold in the millions rather than the hundreds of thousands. This was the successful outcome of the consistent strategy pursued by Martin Knight and the team that prioritised customers capable of offering CML genuine volume. The hope was that, when the supply of SMA OIS was established, others, especially Huawei's domestic competitors, would be encouraged to follow suit.

That CML's SMA actuator had been selected for the P30 Pro soon became common knowledge in the camera module world. Huawei proved unwilling to permit it to be announced publicly. (In any event, Western consumers barely knew the name Huawei. And those who did were unsure how to pronounce it – something that subsequent media attention soon corrected!) TDK, on the other hand, then gearing up to supply the bulk of the actuators used in the P30 Pro, was prepared to promote the technology. CML's director of marketing, Dominic Webber, gave one of the keynote speeches at a TDK Developers Conference held in California in September 2018. He devoted a page of his presentation to the 4-wire OIS actuator produced under licence by TDK. Gratifyingly, at the same conference, there was also a specific promotion of CML's actuator in a presentation by a product marketing director at InvenSense. He referred to SMA as 'the new kid on the OIS block'.

When the P30 range appeared in March 2019, the reviews were overwhelmingly positive. This one, from the well-established US digital media publication, CNET, is representative.

> 'When Huawei first told me about the P30 Pro's four-rear-camera setup, I didn't believe a word. Then I used it. This phone captures zoom and low-light photos unlike anything I've ever seen from a phone … The stabilisation is good, too. When you're zoomed right in on small details, the phone fights hard to keep them steady to help get a crisp shot.' ('Huawei P30 Pro review: The absolute best camera on any phone', Andrew Lanxon, CNET, 17 April 2019)

At the end of June 2019, Huawei announced it had sold ten million units of its P30 and P30 Pro handsets – two full months faster than the P20 range. By September, sales of the P30 Pro on its own had reached 16mn devices. The Mate 30 Pro, which went on sale in that month, encountered a similarly good reception. All in all, over their lifetimes, the two Pro models absorbed 25mn actuators. During 2019, CML's royalty receipts ran at more than twice the budgeted figure. This was obviously an excellent outcome for CML, but it was taking time to set up a smooth-flowing actuator supply chain capable of consistent volume production. Getting TDK and Billu up to speed required a herculean effort, with CML engineers spending weeks or months working alongside local personnel.

It was now confirmed that SMA actuators would be designed into future generations of both P and Mate models. Discussions commenced on a licence agreement that would permit Huawei to upgrade the OIS offered on these from 4-wire to the 8-wire combined AF+OIS actuator then in development. These discussions concluded with an agreement in May 2019. Alone among the major smartphone vendors, Huawei was prepared to take a chance on SMA on terms acceptable to CML. Management could, at last, begin to plan ahead with a degree of confidence. On top of this, the news that Huawei had adopted SMA had engendered much excitement within the industry and every smartphone vendor made contact with CML to learn more about the technology. Huawei's success with the P and Mate 30 and the wider prospects encouraged CML's management to believe the company was firmly on track towards profitability. The September 2019 Shareholder Update projected 2020 revenues of £12mn and an after-tax profit of £1.5mn.

5G AND TRUMP

In the United States, telecommunication cybersecurity issues had dogged Huawei since the early 2000s. In 2012 these came to a head in a US House Intelligence Committee report that, after a year-long investigation, concluded that Huawei acted on behalf of the Chinese government and should not be allowed to operate critical US infrastructure (predominately rural wireless networks). The report highlighted Huawei's ties to the Chinese military and government and the potential for espionage. It also raised concerns over accusations by US businesses that Huawei engaged in the theft of trade secrets and other proprietary data. Huawei, of course, steadfastly maintained it was independent of government and abided strictly by the rules.

During the first half of 2017, as telecommunications companies began testing 5G, US officials again expressed concerns about integrating Huawei equipment into the domestic telephone network. This coincided with the passing of a National Intelligence Law in China, which requires that 'any organization or citizen shall support, assist, and cooperate with state intelligence work according to law'. Taken together, these provoked the Trump administration into action. The government imposed restrictions on Huawei to curb, and then halt, the sale of telecommunications equipment in the United States. Note that these restrictions applied only to telecommunications apparatus. Nevertheless, the negative aura hanging over Huawei as a result of these measures cast a pall over its smartphone activities, just as it was gearing up for a push into the American market spearheaded by the P20.

As part of Huawei's high-profile marketing campaign ahead of the P20 launch, Richard Yu, the CEO of Huawei's consumer division, delivered a keynote speech at the annual Consumer Electronics Show (CES) in Las Vegas in early January 2018. The tone of his speech was downbeat, as this report at the time explains.

> 'The company was widely expected to announce that AT&T would partner to sell the device, a breakthrough for the world's third largest phone maker that is virtually an unknown in the US. Alas, that was not to be … The keynote capped off a tough CES for Huawei, which was riding into the show with what seemed like good news. But then word broke Monday that AT&T was not going to sell the phone. The word came out Verizon may bow out of selling Huawei phones too.
>
> 'The drama, flaring up at one of the tech industry's biggest shows, resurrects the chatter about the security concerns over Huawei's products … At the time, the government said the ban didn't apply to phones, but those concerns have followed Huawei.' ('Huawei CEO: Lack of US carrier support a big loss for you', Roger Cheng, CNET, 9 January 2018)

As we observed in Chapter 6, the US mobile network companies have an outsize influence on the handsets sold in the American market, so the loss of the two biggest operators was a body blow for Huawei. There would be no splashy launch of the P20 in North America. Then, quite separately, hard on the heels of this news, President Trump signed a memorandum applying import tariffs of $50bn to a range of Chinese goods. The reason given was Chinese theft of US intellectual property. Tim Cook's personal relationship with Donald

Trump, allied to pressure from the major American consumer electronics trade association and leading brands, successfully ensured that certain goods including smartphones were excluded.

Until the second half of 2017 the American attitude to Huawei was distrustful but essentially benign. However, by the time of CES in early 2018, the US attitude to China had turned distinctly chilly. Huawei, after success in European markets, was poised to sell its flagship smartphone in the American market. Then, at the very moment this was due to happen, the door closed with a thud. Huawei had become the corporate equivalent of *persona non grata* (*firma non grata*?).

BRITAIN AND HUAWEI

When CML was first introduced to Huawei by Foxconn in 2013, UK–China relations were friendly. In September 2012, a meeting between the prime minister, David Cameron, and Ren Zhengfei, founder and CEO of Huawei, set the tone, with the announcement of a £1.3bn investment in the UK over the next five years. Then, starting in December 2013, when David Cameron visited China on a trade mission and met President Xi Jinping, the mood music accompanying the Sino–British relationship became increasingly melodious.

In September 2015, the Chancellor of the Exchequer, George Osborne, in a speech in Shanghai, talked of a 'golden era' for British–Chinese relations. British business was exhorted to do business in China. So convinced were David Cameron and George Osborne of the merits of their China policy that, early in 2015, the UK – in the teeth of opposition from Washington – announced that Britain would join the Chinese-instituted and led Asian Infrastructure Investment Bank. Later in the year, the government became the first country outside Asia to ratify the articles of agreement of the AIIB. The Sino–British relationship remained cordial but turned sour in 2019 as British politicians expressed their mounting concern over the situation in Hong Kong. In April 2019, it was revealed that the government was looking afresh at the future role of Huawei telecommunications equipment in Britain's 5G network. There was no de facto ban on the sale of Huawei phones in the UK – as there was in the USA – but the company itself had become *firma non grata*.

This is the context in which CML built its relationship with Huawei, during a time when doing business with China was elevated almost to the status of a patriotic duty. The abrupt reversal of UK policy in early 2019 cast a shadow over the forthcoming launch of the P30 range, on which CML had pinned its hopes for real volume – volume that would, via royalty income, propel the company towards financial viability.

19.

Keeping the show on the road 2: management, finances and strategy

R&D AND PRODUCT STRATEGY

As night follows day, the plethora of opportunities that were opening up for SMA led directly to a steep rise in the R&D budget. Serious pursuit of these projects demanded an enhanced engineering resource to work on them. Consequently, as the table below shows, the latter years of the 2010s saw substantial increases in the annual level of R&D spend. Between 2016 and 2018 the total number of employees almost doubled. Income was undeniably on an upward trend but was no match for the uplift in costs. CML therefore continued to make losses. And, inevitably, continued to require regular infusions of cash. The table below (from CML's annual reports) shows the trend.

YEAR	TURNOVER £MN	R&D EXPENDITURE £MN	NEW MONEY RAISED £MN
2015	0.7	2.5	2.5
2016	1.5	2.7	2.0
2017	3.1	3.3	3.0
2018	2.6	4.7	7.5

Operationally, by the end of 2017, a total of 2.5mn SMA actuators had been produced (both AF and OIS) and were out there operating successfully in the field. At the same time, the product range was expanding. CML could now offer:

» Single wire actuators for autofocus only. These were now so cheap that it had become a commodity business, dominated by smaller Chinese VCM producers. However, SMA actuators were competitive in specialised sectors, like drones.
» OIS 4-wire SMA actuators. The 4-wire actuator had matured since 2015 and was now providing the main commercial thrust. Manufacture was carried out under licence by TDK and Billu. The following table, derived from CML's annual reports, shows the rising contribution from 4-wire actuators.

PRODUCT	REVENUE 2015 £000	REVENUE 2016 £000	REVENUE 2017 £000
4-wire actuators (OIS)	655	1265	2762
AF actuators		261	199
Other			116
Total	**655**	**1526**	**3077**

» Finally, high on the agenda at CML was the intriguing prospect – as discussed with Samsung and others in 2016 – of being able to combine the AF and OIS functions in a single unit (the 8-wire actuator), with major perceived advantages in terms of space and cost. But this was still some way off from becoming a commercial proposition.

PLUGGING THE FUNDING GAP

CML continued to raise cash from existing shareholders, culminating in a £3mn convertible loan note facility in the first quarter of 2017. CML had, in fact, secured investment from a new source of funds a few years earlier. Lauriers Investing bought a small holding in 2014. Participation in a subsequent fund-raising substantially increased Lauriers' stake. But the truth of the matter was that, for several years, CML had existed on a restricted diet of irregular small amounts of funding, leaving the company permanently at risk of running out of cash.

By mid-2017, however, it was apparent that the appetite within the existing investor base for additional equity had been exhausted and, to meet its objectives, CML would have to seek further funds from external sources. The result was Project Edison, a comprehensive Information Memorandum prepared by Deloitte in August 2017, directed at outside investors. At this point, the CML shareholding structure looked like this:

Stewart Newton	54.4%
Martin Knight	6.5%
Lauriers Investing	5.7%
Senior management	9.6%
Others	23.8%

The management team at CML remained as it had been since 2012, with Martin Knight as executive chair and Andy Osmant as managing director. There was no CEO so it fell to Martin to provide the leadership the company required. Several industry non-executives were appointed to the board for their expertise.

The first line on the front page of the Project Edison memorandum stated that CML is 'looking to raise £15m of equity funding'. Not all of this would be 'new money', as part of the proceeds would go towards redeeming the convertible loan notes and funding the executive and employee options due to expire early in 2018. The principal purpose of the equity financing was the recruitment of 20 additional staff, 14 of them high-quality engineers from a variety of disciplines. Funds were also needed to expand the company's on-the-ground presence in Asia to provide on-going support for licensees and to maintain contact with customers. With the opening of an office in Shanghai in September 2018, CML completed a network of four bases in the region.

In the summer of 2017 optimism was in the air. Adding up the expected design wins – 11 in all – and projecting out the production of these devices produced a revenue figure of £8.5mn for calendar 2018. If all of this came to pass – admittedly a big 'if' – CML would achieve an operating profit of £1.8mn. In reality, very few of these projects were anywhere near firm. The forecast placed considerable faith in the rapid penetration of the combined AF+OIS 8-wire actuator into design slots at Samsung, Huawei and several other Chinese handset makers.

Attracting the interest of outside investors via the Project Edison Information Memorandum proved to be an uphill task. Valuation was inevitably an issue. Then, in early 2018, the need for additional finance increased as a cash crunch loomed later in the year. By this time, the funding target had been downgraded to £10mn. Deloitte was asked to do a valuation of the company for share-option purposes and came up with a bottom-of-the-range figure of £100mn. This then was the figure that CML used in the 'term sheet' that was distributed to prospective investors. In the event, the full £10mn was raised in two tranches. This top-up equity came in the form of a new class of share: Preferred Ordinary shares, with additional rights attached. The new investors comprised a handful of individuals and some small, specialised funds.

CML's reliance on a drip feed of equity to sustain the business raises a pertinent question: could the company have accessed the larger sum it needed if it had dropped its long-held stance of resisting venture capital funding? VC firms certainly possessed the necessary resources to provide finance in tens of millions of dollars rather than single figure millions. It is doubtful, however, whether CML at this juncture would have been sufficiently attractive to VCs in the absence of a mainstream customer for its devices. CML's hand-to-mouth financing situation also had implications for the company's management structure. Recruiting a CEO of the right calibre was contingent upon the company convincing the candidate that its funding was on a secure footing, then and in the future.

CML's capture of the OIS slot in the Huawei P30 in early 2018 could conceivably have offered the opportunity to approach venture capital firms with a proposition based on a firm contract to supply one of the world's top three smartphone makers. However, as we saw in the previous chapter, the ban on the import of Huawei's telecommunications equipment into the USA had already affected sentiment regarding the company's smartphones, leading in January 2018 to a refusal by the network operators to sell the product. It is a moot point whether the negativity surrounding Huawei would have inhibited US-based venture capital houses from investing in CML, but it is likely. Non-US VCs, untroubled by the association with Huawei, could possibly have provided the funds needed. In any event, the timing was most unfortunate.

Unhappily, the ambitious forecasts in Project Edison were not met. Several of the 11 projects destined to produce, in aggregate, sales of over £8mn in 2018 did not materialise, or the numbers involved were so modest they had little impact. The revenue outcome for 2018 was a disappointing £2.6mn. It will come as no surprise that Samsung was a major factor in the shortfall. As we saw in Chapter 17, the relationship had simply fizzled out. Additionally, after more than a year of interaction, Microsoft decided not to use CML's actuator. Its Surface Duo folding phone eventually emerged in September 2020 to underwhelming reviews. Ubiquiti Labs' FrontRow photo-pendant was a nice idea but, at $400, it was too expensive, and sales were negligible.

20.

Smartphone photography in a multi-cam world

THE MULTI-CAM REVOLUTION

During 2016 and 2017 the smartphone industry was host to a ferment of ideas on how to exploit the potential of multi-camera phones. Every industry participant could see that smartphone photography had reached an inflexion point. For photo enthusiasts, after 20 years in the (single rear camera) wilderness, it felt like the entry into a promised land, where all things were possible.

Until this juncture, improved camera hardware, including more capable actuators, had led to a dramatic improvement in photographic capability. Hardware progress has continued. But, since then, it is software that has been responsible for the heavy lifting in smartphone camera performance, often encapsulated (as discussed earlier in Chapter 10) in the term 'computational photography'. The nature and magnitude of this change was captured precisely in this quote on a specialised photographic website in 2017.

> 'In the era of film, photography was largely related to the recording of light ... But in digital photography, light is converted into bits of data, and increasingly powerful microprocessors and soft-

> ware algorithms allow us to alter that data in ways that would have been impractical or impossible with optics alone. The future of photography is software, and the future has already arrived.' ('How computational photography is changing your conception of photography', PhotoShelter Photographer's Blog, 11 September 2017)

The first dual camera phones were launched in 2016, with Huawei in the vanguard, closely followed by Apple. The primary objective of these was image quality enhancement or, to put it colloquially, better photos. Several possible combinations of modules make up what we conversationally call 'the camera' in a modern phone (sometimes referred to by the chattier members of the techno-commentariat as 'the snapper' or 'the shooter'). The years since 2016 have witnessed a metaphorical explosion in the capabilities of camera modules, working separately or in combination.

As smartphone producers began to explore the increasing number and variety of cameras that a handset could accommodate, they also started looking at functions that led in a different direction, such as using an extra front camera for biometric identification. Biometrics are unique body characteristics used to identify people, exemplified by data derived from fingerprints or facial features. Fingerprint scanning and the two-dimensional use of the front-facing camera were in general use but the level of security they offered was poor. What was needed to enhance these methods was 3D, as the core element of 3D is depth information. Three-dimensional scanning was not new. It had been used for industrial purposes for 50 years. The challenge was to insert an effective 3D sensing module into a package as small as a handset. Fortunately, component technology, particularly lasers and image sensors, had reached the point where this was becoming possible. 3D sensing has applications beyond identification. These are explored later in this chapter.

What does the smartphone camera landscape look like in a multi-cam world? The following list summarises developments during the last decade.

- » The **primary (or main) camera** is the original lens, as used when phones only had a single rear camera. It is commonly a **wide-angle lens**, which can do most things well, and is the default lens for normal use. In terms of its relationship to the additional modules that have emerged, the primary camera can be likened to the sun surrounded by its planets. Some smartphone reviewers contend that these additional cameras are hardly used in practice and, on mid- to low-end phones, are of such poor quality as to be essentially valueless. Clever software on the primary camera, which will have a larger, better sensor, can often perform these functions just as well.
- » An **ultra-wide-angle lens** will, as the name suggests, go further than a standard wide-angle and capture more of a scene. As such, it can be viewed as an adjunct to the primary lens.
- » **Zoom (or telephoto) lenses** are, without doubt, an extremely valuable addition to the smartphone's photographic armoury. (Technically, zoom and telephoto lenses are not exactly the same thing but, for convenience, we will treat them as such.) A crucial distinction exists between optical and digital zoom. Much effort has been expended on trying to replicate the variable (or continuous) **optical zoom** function feature offered by a stand-alone camera. Optical zoom refers to the ability to zoom in on a scene to make an object closer without losing image quality. It is the truest form of image magnification and is achieved by physically moving the lens away from or nearer to the image sensor. Optical zoom is barely possible within the strict confines of a smartphone, which needs to maintain a thin profile with minimal protrusions.

» To get round the sheer physical requirements of optical zoom, the smartphone industry came up with a wholly software-based solution in which the camera 'pretends' to zoom by cropping the image and scaling it to the desired resolution. This is **digital zoom**. The difficulty with digital zoom is that it results in image degradation. Flagship phones offer optical zoom but the fact is that it is not the same as the continuous optical zoom found in dedicated cameras. It mimics optical zoom by seamlessly switching between the lenses, with their different focal lengths. At these focal lengths – typically between five and ten times magnification – it is true optical zoom. Everything in between is achieved by digital zoom. In reality, it is a hybrid solution, albeit an effective one.

» **Periscope (folded) lenses** are another way of inserting the zoom function in a phone by utilising a modern version of an old technology (the submarine periscope) to overcome the limitations of optical zoom as described above. The periscope directs light down the length of the phone. Periscope cameras offer a zoom factor of up to ten times. The following image shows how a periscope lens sends light shifted 90 degrees via a prism mirror system down the body of the phone, where it can pass through multiple lens elements.

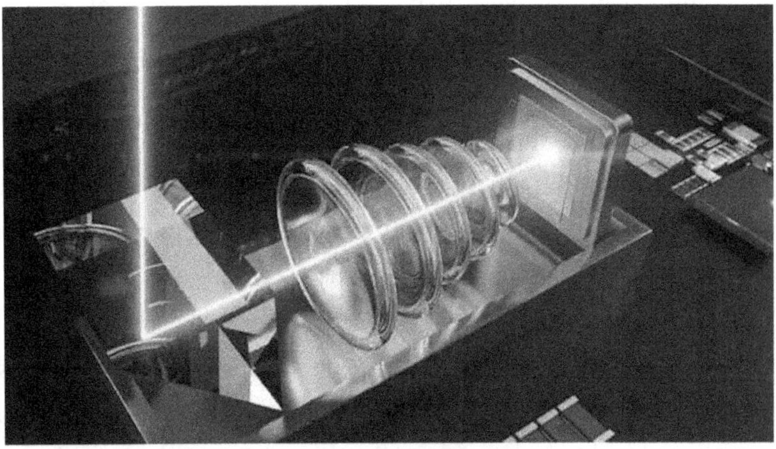

Source: Galaxy S20 Ultra, Samsung Mobile Press, 19 February 2020

The first phone to offer zoom via a periscope telephoto lens was the Huawei P30 Pro. It is a complex and expensive mechanism, so take-up was initially confined to top-of-the-range Android handsets. Apple has been a laggard in this area. One feature of folding optics is that OIS is a must-have on periscope lenses as camera shake is amplified.

» A **monochrome sensor** measures the intensity of light at each pixel more accurately than a colour sensor. It is also more sensitive to light, which is good for low-light situations. In some cameras it can be used for black and white photography, but the monochrome camera essentially exists to enhance the image taken by the primary camera. This is done by blending the sharpness and contrast information it provides with the image reaching the primary camera's colour sensor to create a better overall result. For smartphone cameras to produce colour photos, an RGB (red, green and blue) filter array is placed on top of the sensor. Without this, the camera would only capture black and white images because an image sensor on its own only detects light intensity, not colour.

» A **macro camera** is a lens specifically designed to take close-up shots of small subjects (such as insects). A minority of smartphone cameras have them, but others claim to be able to obtain comparable images using existing camera modules and software. Both Apple and Samsung use the ultra-wide-angle camera to achieve a macro result.

» A **Time-of Flight (ToF)** camera describes the technique this camera uses. A ToF camera creates a 3D map of a scene. It can also be described as a depth sensor camera or a LiDAR camera. Its function is not to take photos but to collect data for computational photography. A ToF camera improves photographs and also provides information for augmented reality applications. Exactly what it does is explained in a later section of this chapter.

Smartphone vendors accordingly have plenty of options when it comes to the rear camera combination. In the early days, amid the excitement that accompanied multi-cam phones, manufacturers were bursting to explore the performance potential of four or five, or even six, cameras per phone. There was a tendency – which peaked in 2021 – to maximise the number of cameras to boost the numbers for marketing purposes. The standard offering has now settled down. Most premium handsets now offer three rear-facing cameras: normally a high-pixel primary accompanied by lower resolution ultra-wide and telephoto cameras. Additional low-resolution sensors (below 15MP) are in decline. The graph below demonstrates the point.

Average number of lenses in smartphones

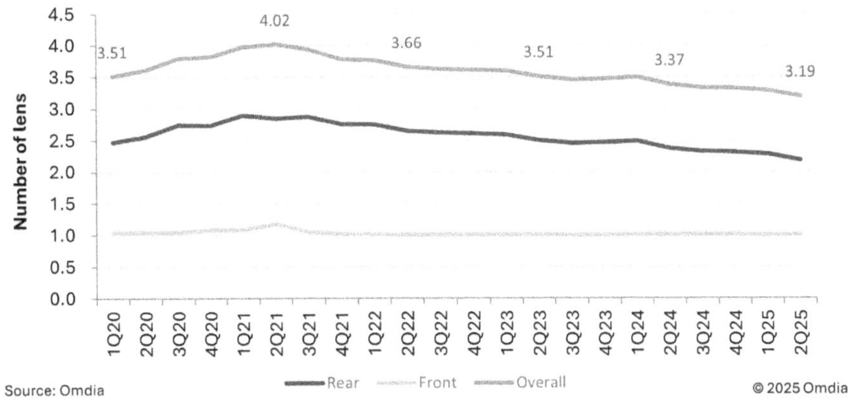

Source: Omdia Smartphone Model Tracker Report, 27 August 2025

For the sake of completeness, we should not forget the humble front-facing (selfie) camera. 'Humble' because smartphone producers initially paid little attention to it. Recently, the growth in video calling and video selfies on social media have spurred a steady improvement in resolution and overall quality. Front cameras now commonly feature 12MP modules. Over the same period, the primary camera has climbed to a pixel count of 48MP or above.

Now for the thorny subject of megapixels. As we noted previously, the number of megapixels a camera offers cannot be simply equated with image quality. The quality of a smartphone's camera depends on a complex interplay of hardware and software. Foremost among these factors is the image sensor. Size really matters. Crudely, the bigger the image sensor, the more light it can receive and the better the image quality. The constraints imposed by the physical dimensions of a smartphone mean that there is a limit to the size of sensor that it can accommodate. Currently, a 1-inch sensor is the limit. In practice, the industry has settled on around 50MP as the optimal size for the main camera on a phone. Handsets with 100 or 200MP primary cameras are made but are in a minority.

Megapixels do matter. Not so much the number but the pixel size. The bigger the pixel, the more light it can capture. Cameras with smaller pixels tend to have issues with digital 'noise', especially in low-lighting conditions. Pixel size is measured in micrometres. Bigger pixels are better. However, powerful hardware and image processing software can make smaller pixel sizes produce first-class images. A phone with excellent processing can often produce a better-looking image from a lower megapixel sensor. To obtain bigger pixels, firms indulge in what is called 'pixel binning', which merges multiple adjacent pixels into larger virtual pixels with higher light sensitivity. The trade-off is low resolution. Without binning, the image can be captured in high resolution as long as the lighting is sufficient.

Unlike taking a shot with a traditional camera, in a smartphone a lot happens beyond – and indeed before – the point of creating the initial image to produce the finished photo. All this is covered by the term 'computational photography'. Nearly every smartphone uses computational photography in some way.

COMPUTATIONAL PHOTOGRAPHY AND ARTIFICIAL INTELLIGENCE

Computational photography is the quiet revolution. 'Quiet' because, in contrast to something tangible and measurable like megapixels or zoom factors or OIS, it is difficult – if not impossible – to enthuse consumers about something as nebulous as software. How do you generate fervour about lines of code? Strangely, given the American penchant for abbreviation, computational photography is not referred to for convenience as 'CP'. As with dumbphones, this book takes a stand on the issue and future references will normally be to CP rather than computational photography in full.

What exactly is computational photography? Digital cameras have used CP since the 1990s when engineers realised that digital images could be manipulated and enhanced using computational techniques. However, CP was only able to realise its full potential in the smartphone, with its powerful and ever-expanding electronic processing capabilities and multiple cameras. A camera digitally captures a photo in two parts: physically, through the actual taking of the photo; and digitally through the application of image processing, which refers to the software that uses computational techniques to enhance the photo. In practice, the two operate side-by-side, prior to, during and after taking a photograph. For example, on opening the camera, software analyses the colour of the local area and detects objects like faces within the scene.

It is believed that the term 'computational photography' was first used in October 2016 when Google introduced its first-generation Pixel phone. It gained a higher profile after Apple highlighted it at the launch event for the iPhone 11 Pro in September 2019. In practice, Apple – like Google – had been employing CP since the iPhone 7 in 2016, which took depth data from its dual cameras to add features to its recently introduced Portrait Mode. A pioneer in this field, Marc Levoy, a professor at Stanford, who spent nine years at Google where he led a research team, defines CP as follows:

> 'Computational imaging techniques enhance or extend the capabilities of digital photography in which the output is an ordinary photograph, but one that could not have been taken by a traditional camera'. ('Computational photography part 1: What is computational photography?', quoted by Vasily Zubarev in DP Review, 3 June 2020)

Where does AI fit into this? In modern smartphone photography, CP and AI are inextricably interwoven and operate in tandem. Both techniques involve manipulating image data and employing algorithms. Both are designed to produce a superior ultimate image. Not that the difference matters in practice. The primary distinction between CP and AI lies in the origin of the image data each uses. In computational photography, the initial image is captured by a smartphone camera with subsequent enhancements and manipulations applied by the device's software. By contrast, AI-generated images are entirely synthesised by algorithms, with no initial photographic input. For all practical purposes, the two technologies should be viewed as a single, synergistic entity – a broad portfolio of techniques for image enhancement.

All this impressive technology raises an important – some might say existential – question: what skill element remains in taking a good photo if it is all done for you? This is an ongoing debate among the photographic *cognoscenti*. Uppermost is the concern that manufacturers will compromise the essence of real image quality by software-centric improvements rather than genuine innovation in camera hardware. Sophisticated smartphones do not just capture an image. They interpret it and reconstruct it, possibly to the point at which it is no longer a record but a simulation.

What does AI actually do? A casual observer of the smartphone scene during the hullabaloo surrounding AI in the second half of 2022 could be forgiven for thinking that phones made before then did not contain AI. They did. In fact, handset photography has benefited from the application of AI since 2016. What happened in 2022 was the emergence, with the launch of ChatGPT, of *generative* AI. Artificial intelligence is based on the concept of machine learning, which allows computers to learn and improve from experience without being explicitly programmed. AI can recognise patterns, interpret complex data and make informed predictions or decisions. It is not new on mobile devices. Siri, Apple's virtual assistant released in 2011, represented a crude form of AI. Apple first introduced machine learning on the iPhone X in 2017. In 2021, Google's Tensor SoCs enabled the Google Pixel 6 to offer machine learning and CP. Voice dictation, language translation and facial recognition are typical applications. AI-powered algorithms can now automatically remove 'noise' from low-light photos, enhance the sharpness of images and create a suitably soft, out-of-focus background for portraits. The camera will often take multiple images, which AI then stitches together to achieve a superior outcome.

Generative AI is different. It is *not* about AI working passively to detect a person in a photo or unlock a phone by raising it to your face. It enables the user to do something wholly new rather than simply building on what already exists, for example, by adjusting a person's expression in a photo. A Google photo-editing feature called Best Take can analyse group selfies and swap facial expressions between images. In other words, Google's AI can create the perfect photo in which everyone is smiling – even if that moment never happened! Generative AI relies on machine learning from vast databases residing in data centres (or on the device itself). Once trained on these databases, it can, for example, recognise objects, scenery or faces. These Large Language Models (LLMs) are therefore 'multimodal', encompassing text, images, audio and video.

Running an AI model on a personal device presents formidable technical challenges. Smartphones lack the computing and energy resources of a data centre and are thus reliant on remote cloud access. This has several disadvantages. First, the exponential growth in AI applications places a strain on cloud capacity. Second, latency (the delay between action and response) can be a problem in certain applications. Third, reliance on a remote server raises privacy and security issues.

As a consequence, the thrust in the smartphone world is towards what is called 'edge computing' or 'on-device computing' – moving the processing of data onto the device itself. Current DRAM memory capacities of up to 12GB on flagship phones are inadequate for AI purposes and need to increase substantially. Due to the burgeoning demand from AI datacentres, DRAM prices are rising, exerting pressure on smartphone costs and therefore prices. Qualcomm especially has moved rapidly in the direction of on-device processing. Its SoC offers Android vendors the ability to embody a host of AI features on their device. Apple is going down the same track. In all likelihood, remote and on-device processing will co-exist, with the cloud handling the more complex tasks.

During 2024, Android smartphone brands heavily promoted the features and services that generative AI could offer the user, focusing on so-called AI Assistants. It is important, though, not to lose sight of the underlying motive for the integration of AI into phones: the desire to create additional sales in a flat market. In June 2026 there was little sign that vendors had met this objective. Counterpoint commented:

> 'GenAI capability has become standard in high-end smartphones... but has yet to give consumers a compelling reason to upgrade. A clear gap remains between devices that are GenAI capable and those on which users regularly engage with AI features. Bridging this gap will require practical, everyday use cases that deliver consistent value' Counterpoint Insight, June 22, 2026.

As if to make the point, the iPhone 17 series sold well in the final quarter of 2025 despite a limited AI offering compared to its rivals.

THE ASCENT OF THE PHOTO CHIP

The semiconductor industry has responded to the ever-increasing demand from the mobile phone manufacturers for additional CP and AI processing capacity by developing specialised chips, some of which were directed at photographic tasks. 'Photo chips' can be separate ICs or are fabricated as a constituent part of the silicon real estate that is the SoC, with advantages in cost, space and power consumption.

The essential function performed by an SoC was described in Chapter 7. Traditionally it comprises a Central Processing Unit and a Graphics Processing Unit, these being the two key functions. A modern smartphone SoC, however, often packs far more than this. The usual elements are:

- **Central Processing Unit (CPU)**. The 'brains' of the SoC. In essence, it runs the code for the operating system and the apps.
- **Graphics Processing Unit (GPU)**. Deals with graphics-related tasks, such as gaming. GPUs are designed to operate much faster than a CPU.
- **Neural Processing Unit (NPU)**. Used in phones to accelerate machine learning activity. Specific to AI applications.

» **Image Signal Processor (ISP)**. Converts the data from the phone's camera into image and video files.
» **Digital Signal Processor**. Handles more mathematically intensive functions than a CPU, for example, analysing gyroscope sensor data.
» **Video encoder/decoder**. Performs the conversion of video files and formats.
» **Modem**. Provides access to the network carrier.

Computing in the 1990s was simple. It consisted of a Central Processing Unit (CPU) and memory. Then a new kind of processor, the Graphics Processing Unit (GPU), began to attract attention, driven by the consumer desire (on personal computers) for graphic-intensive games. Graphics, especially 3D graphics, needed something more than a CPU. While a CPU is designed to tackle calculations sequentially, a GPU is capable of performing millions of calculations in parallel.

Nvidia, founded in 1993, was early into this market, as were many others. By the mid-2000s, it had forged ahead to become the undisputed leader in GPUs for gaming. From this base, Nvidia progressed to the overwhelming share it holds in GPUs for training large AI models. Its position owes as much to the software that accompanies the chip as it does to the ingenuity of its designs. CUDA (Compute Unified Device Architecture), a software ecosystem the company created in 2006, has become a critical part of AI infrastructure. More than three million developers use CUDA, which will only work with Nvidia GPUs. Nvidia has come to dominate AI chips, just as it originally attained leadership in graphics for gaming. Nvidia regards itself as being in the business of accelerated computing, now widely recognised as referring to a computing approach in which calculations are performed on specialised processors to achieve faster execution.

AI was initially held back by the lack of computer power to make deep learning models possible. Nvidia – and others – perceived that

GPUs opened the door to AI. Machine learning lies at the heart of AI and the parallel processing capability offered by GPUs proved to be faster and more efficient at training or running machine learning models. Nvidia's AI chips supplement an existing CPU-based set-up, operating as accelerators, with an ability to churn through the most intense workloads.

Neural processing refers to the way the human brain works. Semiconductor advances offered the opportunity to replicate this process. It soon became apparent that machine learning could be accomplished even more effectively with a dedicated processor, working in conjunction with the CPU and the GPU. This became known as a Neural Processing Unit. An NPU was better at machine learning tasks than a GPU and consumed less power. Apple, Google, Nvidia and numerous others, including Qualcomm, Huawei, Samsung, Oppo and MediaTek, have designed NPUs for their own requirements.

Neural Processing Units are the latest specialised processors to find their way into phone SoCs and enable on-device machine learning. Apple was early into this area with its own customised chip, which it calls its Neural Engine, introduced in 2017. Apple's early and continued emphasis on software has enabled the iPhone to maintain its reputation for photographic excellence while eschewing cutting-edge developments in camera hardware. The Neural Engine handles the algorithms behind AI, augmented reality and machine learning. It helps with a variety of tasks, including photography (working in conjunction with the Image Signal Processor, see below), face recognition, voice recognition, image search and shutting down unused apps to save power.

An Image Signal Processor (ISP) is a specialised Digital Signal Processor that processes the electric signal from the image sensor and rebuilds the image to ensure the final photo is as flawless as possible. ISPs enable many user functions that have greatly enhanced photography and video, including automatic exposure control, automatic focus and object detection. Android vendors naturally wish to find ways of distinguishing their handsets from the pack. ISPs are one of the few smartphone silicon components that can be moved outside the SoC with relative ease. There is therefore a move by smartphone manufacturers to develop custom imaging chips rather than rely on the built-in ISP that Qualcomm and MediaTek offer in their SoCs.

Google's own Tensor brand SoC incorporates a self-designed NPU. Developed specifically to power Google's Pixel 6 smartphone, in place of Qualcomm, Tensor has not been an outright success. Every generation gives the impression of being consistently one step behind the equivalent Snapdragon in terms of performance. Even a company possessing the resources Google can command, finds it hard to emulate Apple in creating a cutting-edge custom silicon mobile phone processor.

THE SMARTPHONE PHOTOGRAPHER'S ARBITER

In the modern world, every corporate entity is the subject of continual assessment. Typically, in a business sector, a few organisations emerge as arbiters of competitive performance, accepted by all concerned as independent and impartial. In some instances, just one becomes recognised as *the* authority by all participants. Photography – and not just smartphone photography – has such an organisation called DXOMARK.

Founded in Paris in 2003, DXOMARK is universally acknowledged to be the ultimate judge of camera quality and performance. Cameras are evaluated against a wide set of criteria to arrive at a 'Top Smartphones by Camera score'. The database is searchable. In the words of the company's website:

> 'DXOMARK is a French engineering services company whose testing, software, and benchmarking services help smartphone, speaker and digital camera makers create the best possible products.'

DXOMARK employs 120 people of whom 100 are scientists and engineers. The firm evaluates photo and video quality separately and then combines them into an overall score. On its website, a DXOMARK camera league table looks like this:

Top Smartphones
by Camera score

Sort by Camera

Filter by Segment ^ Brand ^ Bar/Fold ^ Launch Price ^ 2013-2026 ^ Reset all

Rank	Device	Launch Price	Launch Date	CAMERA >	SELFIE	AUDIO	DISPLAY	BATTERY
1.	Huawei Pura 80 Ultra	$1199	Jun 2025	175 V6	-	-	-	-
2.	Vivo X300 Pro	$1299	Oct 2025	171 V6	-	-	-	-
3.	Oppo Find X8 Ultra	$899	Apr 2025	168 V6	-	-	-	-
=	Apple iPhone 17 Pro	$1099	Sep 2025	168 V6	154	-	-	-
5.	Vivo X200 Ultra	$1100	Apr 2025	167 V6	-	-	149	-
6.	Oppo Find X9 Pro	$1299	Nov 2025	166 V6	-	-	-	-
=	Xiaomi 17 Ultra	$1299	Dec 2025	166 V6	-	-	-	-
8.	Honor Magic8 Pro	$1299	Oct 2025	164 V6	-	-	-	-
=	Motorola Razr Fold	$1999	Mar 2026	164 V6	-	-	-	-
=	Motorola Signature	$999	Jan 2026	164 V6	-	-	-	-

Source: DXOMARK (dxomark.com)

These scores are guaranteed to set camera module industry pulses racing, as evidenced by the frequency with which the techno-commentariat refers to them. For example, this reaction to the Huawei Mate 30 Pro.

> 'The boffins in the DXOMark labs have been busy putting the Huawei Mate 30 Pro's new camera through its paces. Surprise surprise, Huawei's latest huge 40MP camera sensor regains the crown for the Chinese manufacturer. In total, the Mate 30 Pro

> scores 121 points in DXOMark's camera test, a full four points higher than the Galaxy Note 10 Plus 5G. ('HUAWEI Mate 30 Pro camera reclaims the DxOMark photography crown', Robert Triggs, Android Authority, 26 September 2019)

Smartphone vendors frequently quote the DXOMARK score their latest flagship model has achieved – assuming, of course, it is at or near the top of the list – in their marketing material.

VIDEOGRAPHY RISES

Post the multi-cam revolution, smartphone photography attained new heights. Shooting video, however, remained a secondary activity compared with taking stills. This was despite many high-end phones featuring a 4K high-definition (HD) video (4K refers to a video resolution that is approximately 4,000 pixels wide. 4K video has four times the resolution of standard HD video). However, when it came to what we now call videography, people generally still turned to digital movie cameras or camcorders. That is until 2020, the year that changed everything. The pandemic coincided with – and fed – a boom

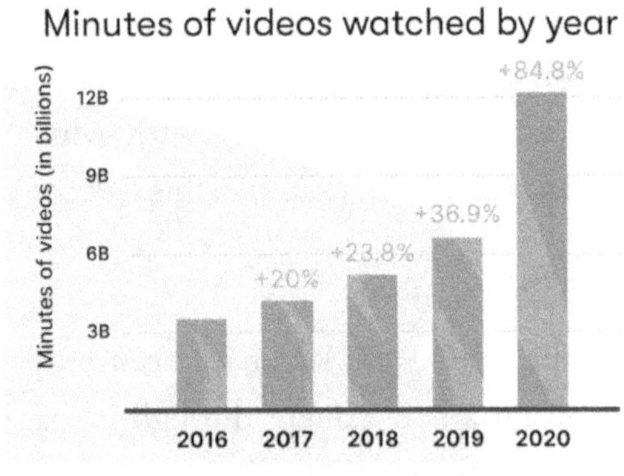

Source: 'Content Distribution: The Marketing Executive's Guide', Terakeet, 22 October 2021

in video. The graph below uses data from Wistia, a video management and hosting company, to show the leap in video consumption that occurred in 2020.

Since 2020 video viewing on social media platforms such as YouTube, Instagram and TikTok has exploded, for both social and commercial purposes, leading to a step function increase in demand for content creation. Fortunately, advances in both hardware and software had, by this time, greatly enhanced the capabilities of 4K video in handsets.

The new breed of non-professional content creators that emerged during the pandemic generally work from home and use the smartphone camera they already have rather than buying a digital camera just to shoot videos. Full-length feature films can now be successfully shot and edited on a phone. Video marketing by vloggers (bloggers who use the medium of video to get their message across) has grown exponentially, as businesses discovered its value in driving sales, to young people especially. Now 5G, with its greater capacity and faster speeds, is giving a further fillip to videography. For content creators, 5G offers the ability to upload high-quality video in seconds rather than minutes or hours.

3D TECHNOLOGY ON THE FRONT

In 2017 Apple surprised the smartphone world by introducing a new method of biometric user authentication called FaceID on the front of the iPhone. Until then it had relied on a fingerprint scanner, branded TouchID, to do the same job. Fingerprint scanners had become standard across the mobile world by this time. Generally, they worked well. Some manufacturers opted for 2D face scanning for identification, while recognising that it was inadequate for security purposes. Apple's FaceID is more sophisticated and secure, employing its proprietary TrueDepth front-facing camera system. Having first taken a 2D image, a structured-light transmitter shoots out 30,000 infrared dots to create a 3D image of your face for matching purposes utilising software algorithms. Structured light is one of several techniques used to acquire the data needed for 3D.

If the objective is to obtain useful information for security and identification purposes the correct term is '3D sensing' rather than '3D imaging', though the two are often used interchangeably. 3D sensing is the process of obtaining length, width and depth information electronically. Its primary function is detection, not image creation. In 3D imaging the objective is the construction of a computerised image of whatever has been scanned.

Just as we have a tendency to talk about *the* camera in a phone, so people refer to *the* TrueDepth camera. In reality, it is a complex subsystem that sits at the top of an iPhone screen, as per the image below. The system relies on a special type of laser, called a dot projector, to cast the necessary infrared beam on the face being scanned. Creating a supply chain for its inclusion in the iPhone was a major task. The total cost of FaceID must have run into the low billions of dollars, setting a high bar for its handset rivals.

How did the Android community react to this step change in mobile phone authentication? In 2018 the technological pundits forecast that the Android makers would follow suit. They did not. By and large, the individual companies decided to stick to fingerprint sens-

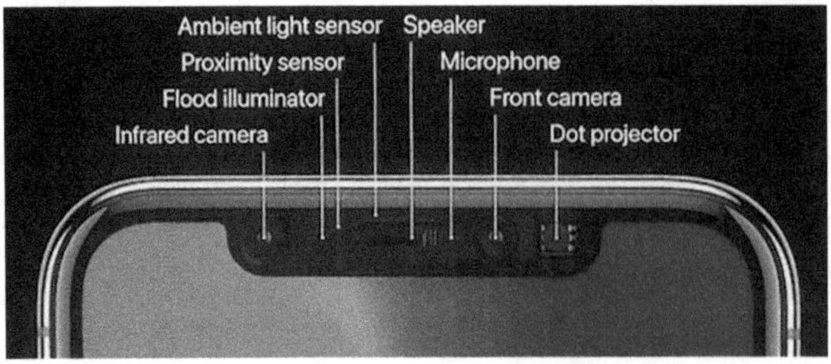

Source: iPhoneFAQ.org

ing, heading in the direction of placing the sensor under the screen to avoid taking up screen space. Vivo introduced the first under-display scanner in 2018. This has since become mainstream on Android flagship phones. FaceID was not ignored. Attempts were made to replicate what Apple had done but these were beset with issues. For one thing, Apple holds several key patents in this area. It was also a struggle to access suitable lasers. On top of this, the Apple authentication solution was – unsurprisingly – very expensive and not one that price-conscious Android vendors could easily tolerate.

Some Android manufacturers, notably Huawei, Google and Honor, replicated FaceID in some of their top-end models but, broadly, firms have settled for less secure face authentication that simply uses the front-facing camera to capture a 2D image. Generally, everyone other than Apple has continued to place their faith in the tried and trusted technology of fingerprint scanning, often in conjunction with face scanning.

3D TECHNOLOGY ON THE BACK

Meanwhile, 3D was advancing on the back of the phone. By 2019 Android manufacturers were launching handsets with rear depth sensors in a few high-end models. These sensors do the same (3D) job as FaceID but for a different purpose: to assist the camera, so enhancing photographic performance. Instead of structured light, these rear depth cameras used a less costly alternative scanning technique called Time-of-Flight, or ToF for short. ToF technology has characteristics that make it far more suitable for rear-facing camera applications than structured light. Range is a critical difference: ToF will detect objects up to 5 metres away compared with an optimal range of 0.5–1.5 metres for structured light. The trade-off is inferior resolution.

ToF Android handsets became available during 2019 although there was no groundswell of enthusiasm for the technology. 3D sensing was sold to the consumer as adding a dimension to smartphone photography. It would, for instance, help the camera to decide whether to expose a shadowed person in the foreground rather than a bright landscape behind. There were persistent rumours that Apple would install a 3D camera on the rear of the iPhone. This technology eventually saw the light of day in October 2020, in the iPhone12 Pro. The Apple press release said:

> 'An all-new LiDAR Scanner comes to the Pro line, offering the ability to measure light distance and use pixel depth information of a scene. This technology delivers faster, more realistic AR experiences and improves autofocus by 6x in low-light scenes for more accuracy and reduced capture time in photos and videos.'

The arrival of LiDAR, and the heavy marketing campaign associated with it, pushed ToF, and its potential for the delivery of augmented reality on phones, several notches up the scale of public consciousness. Since then, iPhone Pro or Pro Max models have had a – minimally promoted – LiDAR sensor incorporated in the rear camera system.

LiDAR stands for light detection and ranging. It uses lasers to ping off objects and return to the source, measuring distance by timing the flight of the light pulse. LiDAR is a *form* of Time-of-Flight. However, because it has been so assiduously promoted by Apple, the smartphone-consuming public often gains the impression that it is synonymous with Time-of-Flight. It is not. LiDAR differs from standard Time-of-Flight in the nature of the waves that are used to construct a 3D map or image. It may be new to mobile, but LiDAR has been extensively used in other settings, such as creating accurate topological maps, for many years. Consequently, the two alternative ToF sensing methods are:

» **Direct Time-of-Flight** (aka LiDAR) uses short pulses of infrared light to detect the position of an object, measuring the time difference between send and return.
» **Indirect Time-of-Flight** sends high-frequency modulated infrared light that returns in a different phase. It therefore uses the phase, rather than the time signature, of the reflected light to measure distance.

Direct ToF (LiDAR) is expensive due to its intrinsic complexity and the difficulty of inserting it into a mobile device. Conversely, indirect ToF can be made more cheaply and has advantages in 3D imaging. Time-of-Flight modules have a wide range of applications beyond smartphones. Major growth is occurring in automotive (for cameras) and in the home (for products such as robot vacuum cleaners).

AUGMENTED REALITY (AR)

While the initial impetus for installing Time-of-Flight was the desire to boost a phone's photographic capabilities, there was an additional motivation. Within the smartphone industry, considerable excitement was building up around ToF as a way of stimulating greater usage of augmented reality. AR is the technology of superimposing digital objects on your view of the physical world. It is similar to, but distinct from, virtual reality (VR), which immerses the wearer in an entirely internal world. Proponents argued that, if AR became popular with consumers, it would give a fillip to handset sales.

AR on phones had been possible for some years using clever software. In June 2017, three years before LiDAR appeared on the iPhone, in an effort to kick start AR, Apple announced its ARKit development platform. Apple developers were exhorted to 'Build unparalleled augmented reality experiences for hundreds of millions of users on iOS'. Tim Cook has been a consistent apostle of AR, speaking publicly to this effect on numerous occasions. A typical statement is this one, made in 2016.

> 'I do think that a significant portion of the population of developed countries, and eventually all countries, will have AR experiences every day, almost like eating three meals a day. It will become that much a part of you … AR is going to take a while, because there are some really hard technology challenges there.' (Panel discussion at a conference in Utah. Quoted in Macworld, 3 October 2016)

Google followed suit in 2018 with ARCore, a software development kit that allowed Android developers to create augmented reality apps. Even before these development tools became available, in 2015, Snap, the owner of Snapchat, the social media platform popular among young people, picked up AR and ran with it to create what it called Lenses. Using the front-facing camera in the app, Snapchat's mapping technology could register your face, render a 3D model and drape a digital image over it, transforming you into a dog or a zombie, for example. It was there for fun. Snapchat Lenses was an instant hit among its user community.

ARKit for iPhone and ARCore for Android required no special sensors, relying on the camera and the motion sensors in the phone to create an AR experience. Characteristically, in Google's case, software does more of the work. Modern multi-cam phones can generate depth map data in different ways: using ToF/LiDAR sensors, dual camera setups or by taking two pictures with a single camera and applying intelligent algorithms. Google responded to the introduction of LiDAR with an extension to ARCore called Depth API, allowing developers to create a depth map using a single camera.

A major AR event occurred in 2016 when Niantic, a software development company spun out of Google the year before, launched its app Pokémon GO internationally. The game overlaid augmented reality objects on what could be seen through mobile device cameras. The idea was to locate, capture, train and battle virtual creatures, called Pokémon GO, which appear as if they are in the player's real-world location. It instantly became a worldwide phenomenon, reaching a peak of 130mn users in 2018. There has been nothing like it since, though not for want of trying on the part of Niantic. Additional applications did emerge. Unlike VR, which requires a headset, AR mobile games can be played on any modern smartphone. Gaming on phones has indeed gained popularity

Online retail has long been viewed as a major growth opportunity for AR, as it allows purchasers to customise and virtually 'try on' shoes and clothing and 'place' objects in a chosen environment. For example, IKEA immediately perceived the potential of AR for remote shopping for furniture and, by September 2017, had announced an app called IKEA Place, built on Apple's ARKit. This allowed a would-be purchaser to 'see' how a sofa or coffee table would look in a room in 3D and true to scale. However, for AR to really take off, developers needed to create apps with broad appeal like Pokémon GO. All the elements were in place for take-off, but it needed an app catalyst, preferably several catalysts.

A ToF/LiDAR sensor is better at measuring distances and recognising surfaces than software-based AR and the response is immediate. The thinking in 2020 was that, led by LiDAR, apps created for augmented reality in the consumer sphere would power the inclusion of ToF sensors in smartphones by offering a superior AR experience and, in doing so, boost smartphone usage and demand.

21.

Google enters the fray

GOOGLE AND ANDROID

Google did not take the idea of competing in handsets seriously until the multi-cam revolution, when it believed it had a one-time opportunity to bring its software expertise to bear on smartphone photography. The Google–Android relationship is a crucial element in the competitiveness of the Android phone manufacturers in their battle against Apple. The following paragraphs explain how the Google–Android relationship works. The headline fact that Android is open source and grants Android licences at no cost to allcomers may convey an impression of benevolence and a laissez-faire attitude towards its licensees. This is not the case. It is a hard-headed, commercial calculation – and a successful one at that. For Google, the Android operating system opens the door to a lucrative income stream over which it retains firm control.

In 2025 Google Services – the division that powers Alphabet, as the listed company is now called – reported a generous 41 per cent operating profit margin on revenues of $343bn. Within Google Services, two-thirds of this revenue comes from 'Google Search & Other'. Alphabet does not divulge the profitability of Search. At root, this is what drives Google. The Google Play Store also falls within Google Services. Though Google Play Store is nothing like as significant a contributor as Search, court filings in an anti-trust case relating to 2019 afford a glimpse into its profitability. In that year the Google Play Store earned a profit of $8.5bn on revenues of $11.2bn – a 76 per cent margin. Hosting apps is a lucrative business, as a deep dive into Apple's accounts – were that possible – would no doubt testify.

The core Android operating system, AOSP, is free and open source. Most devices run on the version developed by Google, Google Mobile Services (GMS), which comes bundled with Google Play Store and other core Google services. Smartphone users in most parts of the world rely heavily on Google-generated apps and generally prefer to continue to use them on a new device. The GMS version of Android is free to licensees. It is updated annually by Google. Smartphone vendors can, of course, add their own skins to the software they receive from Google.

Google Mobile Services exerts a powerful grip on the Android ecosystem. Amazon launched its own Fire Phone handset in 2014 and determined to create its own app store – which still exists – in place of Google apps. But consumers reacted adversely to the proprietary operating system and the absence of familiar apps and it was discontinued a year after launch. As we observed in Chapter 16, Huawei is in the same position as Amazon was then – though not by choice. US restrictions on the company's access to Google software means it cannot offer the standard Google apps on the handsets it sells. In China this has no practical effect but the absence of Google software on Huawei handsets seriously inhibits sales outside China. It is possible to download Google apps onto a Huawei phone after purchase via its own version of the Google Play Store, the AppGallery, using various workarounds, but only a minority of buyers will bother.

A smartphone manufacturer wanting to use GMS does not need to pay Google anything. This gesture on the part of Google should not be mistaken for altruism. It is not. A company wanting to use the OS has to sign a document called a Mobile Application Distribution

Agreement to be able to use the full version of Android with the Google Play Store and other Google apps. (This agreement goes by the acronym, MADA, which, given Google's vested interest in maximising the purchase of apps, could just as well stand for Make America Download Again.) MADA insists that a smartphone licensee:

- Offers purchasers of its device the entire set of Google apps (YouTube, Gmail, Maps, etc. and especially Search and Chrome), all of which must be pre-loaded on the phone.
- Pre-installs Google Search as the default search engine.
- Ensures Google apps are prominently displayed on the home screen.
- Applies these rules to its entire range of devices.

Android is indeed free if the licensee accepts the full Google package but it comes with a series of obligations intended to benefit Google.

THE GOOGLE MONEY MACHINE

How exactly does Google generate income from Android? The short answer is Google Search. Search is the big money-spinner. It makes money primarily through advertising. Advertisers pay to display sponsored results when users search for specific keywords. Paid search generates revenues for Google via the number of clicks these ads receive, in a business model known as Pay-Per-Click. Search engines additionally track user data and then use these data points to improve the targeting of advertisements, not just those on the search engine site but also on other websites and apps.

A further source of revenue is the Google Play Store. It contains popular Google apps and acts as a distribution channel for many more from third-party developers. It is estimated that 90 per cent of

Android apps are downloaded from the Google Play Store. The vast majority are free, but some are paid for. All apps hosted by Google Play must pay a small up-front fee. Google earns a hefty commission when paid-for apps are downloaded. It benefits indirectly from free app downloads via advertising and commission on in-app purchases and subscription fees.

Google Search is a de facto monopoly. Its share of the global search engine market hovers between 85 per cent and 90 per cent. It is the default search engine on nearly all Android devices. Users are free to select an alternative search engine but, as ever, inertia rules, and few do. What is less well known is that Google is also the default search engine on iPhones. What purports to be Apple's Safari browser search engine is, in fact, Google Search. An anti-trust suit filed in California in 2022 alleged that more than half of Google's search business was conducted through Apple devices, presumably reflecting the iPhone's wealthier customer base.

As in Android, it is easy to change the default search engine in iOS – there are four alternatives – but few users avail themselves of the opportunity. True, many users would actively choose Google anyway on the basis of a superior search experience, but this does not justify the effective denial of choice at the point of purchase. To ensure it retains pole position, Google pays Apple a large sum of money every year, reportedly $20bn. Samsung too is the recipient of Google's 'largesse', as are other companies. These payments to third parties to drive traffic towards Google Search are recorded in the Alphabet annual report under the anodyne heading 'Traffic Acquisition Costs'.

In 2024 Google came under anti-trust pressure, with the US Department of Justice alleging the company struck anti-competitive deals with Apple and others for prime placement of its search engine. Admittedly, for Google, the emergence of AI raises the spectre of Search being undermined by AI-driven alternatives. To counter this, Google has embedded its Gemini multimodal AI chatbot into Search. As of late 2025, a judicial ruling – recognising the existence of rising AI-based competition – merely insisted that Google modify the search engine distribution arrangements it has in place.

THE PLUNGE INTO HARDWARE

In 2016 Google, despite having no experience of hardware, decided to become fully involved in the handset market. Its Pixel-branded phones are manufactured under contract by Foxconn. Part of the motivation for getting into handsets was the belief that these would act as a 'reference' product, showing its licensees what could be done with software to enhance Android's appeal. As we have noted on several occasions throughout this book, there is an unwritten rule in business that suppliers should avoid competing with their customers. Google though, as a licensor of the technology, is in a less exposed position. In a broad sense, it is a supplier of that technology but, unlike a component that is at risk of being supplanted by a competitor's product, its licensees are tied into the software it provides. For licensees, moving to another platform is extremely difficult. Google therefore has considerable leverage over them.

However, the fact is that, for Google, making a mark in smartphones has been a struggle. The company ships around ten million Pixels a year, miniscule in the context of 1.2bn handsets sold worldwide annually. Pixel phones are aimed at the mid and upper end of the market, competing primarily with Apple and the Samsung Galaxy S range. The premium handset orientated American market should be fertile ground. Chinese devices are effectively shut out of the US market, allowing Google to take the fight directly to Apple and Samsung, with their combined 80 per cent share. Yet Google accounts for fewer than four million devices a year out of around

125mn. In terms of market share, Pixel is down in the undergrowth, well below Motorola (Lenovo) which benefits from its longstanding name recognition.

Part of the motivation for going into hardware was to allow the phone to act as a poster child for what could be achieved in Android through the application of its formidable software skills, with the focus on computational photography (CP). In this respect Google has achieved its objective: its CP capabilities are considered to be on a par with Apple's. Pixel phones have acquired a reputation for excellent software allied to mediocre hardware. Why have they made so little impact commercially after several years in the business? There are a host of reasons. Here are just some of them:

» Google's lack of experience in developing and manufacturing hardware. Software is easy to adapt and change whereas hardware has to be frozen in time.
» Being late to the party – other brands were already embedded in the consumer's psyche.
» Lack of impetus for fear of pushback from its licensees.
» Few resources devoted to marketing/no retail footprint.
» Absence of carrier partnerships, especially in the USA.
» Emphasis on software-intensive applications, especially the camera experience, offset by a less-than-appealing design and some quality and reliability issues.
» Adherence to its own Tensor SoC since 2021, with shortfalls in performance relative to its Qualcomm peers.
» The fact that Google's own skin, Pixel UI, originally stuck close to stock Android, with limited extra features (but has since taken advantage of its ownership of Android through improved integration of software and hardware).

In 2021, Google, having learnt much since 2016, decided to pursue a more aggressive strategy with Pixel. This became flesh in 2022 with the launch of the Pixel 6, followed later in the year by the Pixel 7 range. These handsets were generally well reviewed, with the usual praise for the camera. In February 2023 the CEO of Alphabet hailed the 2022 Pixel range as 'the best-selling generation of phones the company has ever launched'. Undeniably so, but not a high bar to overcome!

All of this gives rise to an intriguing question: why does Google bother at all with the handset market? It is a software powerhouse and becoming involved in hardware is a departure from the company's origins and corpus of expertise. Pixel revenue is not material to the company as a whole – at a guess less than 2 per cent of group turnover in 2025. And it is almost certainly loss-making.

The substantive issue here is the relationship with Samsung. It is by far the dominant seller of Android handsets. Its premium Galaxy line sells well in higher income countries, precisely where Pixel handsets have appeal. In Europe and markets like Japan and Australia, Samsung and Pixel meet head-on, albeit with competition from Chinese vendors. For Samsung and Google, co-dependency risks leading to conflict. The two companies have a long history of rivalry. Samsung is the main window through which Google reaches the public. Equally, Samsung needs Google's software to be effective in the marketplace.

Early in the history of smartphones, Samsung twice tried to get out from under Google by launching its own operating system. Neither time did it catch on. Pixel can be viewed as a kind of insurance policy in case Samsung ever decided to go down the path of creating its own OS. Competing directly with Samsung – but not too much – is a delicate balancing act. Following much muscle-flexing, the two firms seem to have arrived at a *modus vivendi.* The emergence of 'Agentic AI' and AI-based search opens up the possibility that Samsung will turn to a third-party provider for these services.

One important way in which possible friction has been diffused relates to the Pixel's manufacture. Assembled by Foxconn, the phone contains a substantial number of Samsung parts. According to an analysis by Counterpoint Research, just over half the value of the components in the Pixel 7 in 2022 could be attributed to Samsung. The proportion of Samsung made parts will have declined since then, if only because Google has switched the fabrication of its Tensor SoC from Samsung's foundry to TSMC. Nevertheless, every Pixel sale benefits Samsung directly.

The deal on Pixel between Google and Samsung is part of a multi-faceted relationship between the two companies that includes the creation and production of augmented reality glasses and headsets. These projects complement the software ecosystem it is developing, called Android XR, to facilitate compatibility between devices. Chapter 24 covers the brave new world of extended reality.

PART 4

RETREAT AND REGENERATION

22.

SMA makes progress, geopolitical clouds gather, 3D beckons

VOLUME AT LAST

Chapter 18 left CML, in the second half of 2018, readying itself to supply 4-wire SMA actuators for the soon-to-be-launched Huawei P30 Pro. High hopes were placed on the P30 and the increasing prospect of a broader relationship with Huawei. Discussions with Huawei on adopting the 8-wire combined AF and OIS actuator across the range had accelerated. There was much rejoicing when TDK (with Billu as a second source) delivered 400,000 actuators for the forthcoming P30 Pro during December 2018.

CML had the wind in its sails. Huawei was gaining ground everywhere around the world – except the USA – and was firmly on track to make 240mn handsets during 2019. This would place Huawei behind Samsung (300mn units) and ahead of Apple (200mn units). The P30 series sold well, reaching ten million sales in just 85 days – 2 full months faster than the P20. The year saw several further positive developments.

- In May 2019, CML signed a long-term agreement with Huawei for the supply of 8-wire actuators for its flagship P and Mate models. Huawei would continue to insert 4-wire units until the cheaper combined AF+OIS actuator was ready. CML also achieved a design win for a model for Huawei's sub-brand, Honor.
- TDK had ramped up production of the 4-wire actuator for Huawei. Billu, on the other hand, was taking time to get up to capacity.
- Chinese smartphone vendors expressed interest in 4-wire SMA but, as ever, wanted to wait and see it working at Huawei before making a decision. Xiaomi was in the vanguard here. New Shicoh, a CML licensee, was its preferred actuator maker. Unfortunately, New Shicoh, like Billu, was proving slow to get into production.
- Huawei was keen to establish a third source of SMA actuator supply, so discussions were initiated between CML and Alps Alpine, which as we saw earlier, was one of Apple's two actuator suppliers. These progressed swiftly and Alps moved to establish SMA production capacity for 2020.
- Progress was made in haptics and 3D sensing (see later in this chapter).
- At the end of the year, shareholders were told that both Samsung and Apple had indicated interest in adopting the existing 4-wire SMA OIS actuator.

In 2019 CML delivered 24mn OIS actuators. The great majority went to Huawei, but several other smartphone producers absorbed small quantities. This fresh momentum was reflected in the financial results. Revenues were £6.6mn, well up on the previous year. The chairman's statement, written in January 2020, was distinctly upbeat:

> '2019 has been an impressive year in the development of Cambridge Mechatronics Limited (CML). The company now

> has nine licensees, from which substantial licence fees have been received; and for the first time saw adoption of its 4-wire OIS actuator in multiple smartphones. For the future, of significance was the licensing in 2019 of CML's next camera designed product, the 8-wire actuator. We have high hopes that this product will be very attractive in the market, with the prospect of significant volume of sales in 2020 and beyond.'

Unfortunately for CML, geopolitics were beginning to cast a long shadow over the company's prospects.

FROM HERE TO ENTITY

The years 2013–19 were good ones for Huawei. However, by December 2019, the solid gains achieved against Samsung and Apple had started to fade. The company's global smartphone market share (according to the tech consulting firm, Canalys) fell from 16.6 per cent in the final quarter of 2018 to 15.2 per cent in the last quarter of 2019.

What happened? In a word (strictly, two words), the US Entity List. Huawei was placed on this list on 16 May 2019. This move by the Trump administration should be seen in the context of the pressure already exerted by the USA on Huawei's 5G equipment and the arrest, in December 2018, of the Huawei executive, Meng Wanzhou, the daughter of the founder. The allegations against her related to the violation of US sanctions against Iran. The so-called Entity List banned Huawei from acquiring components or technology (including software) from US companies without government approval – which, barring some temporary reprieves, was not forthcoming.

For a time, there was some optimism on the part of Huawei that the differences between the USA and China could be resolved. It remained conceivable that the Trump administration could 'turn on a dime' and reverse current policy. Huawei derived encouragement from the tale of ZTE, another Chinese telecommunications manufacturer with a smartphone arm. ZTE was prohibited from buying US-origin products in April 2018. These restrictions had a profound impact on ZTE's business but the ban only lasted three months. ZTE survives and remains a minor competitor in smartphones. But for Huawei there was no respite.

The Entity List is a curiously named directive administered by an obscure branch of the US Department of Commerce, the Bureau of Industry and Security. The BIS is tiny in the context of the Commerce Department, employing 350 agents and officers who monitor trillions of dollars of transactions around the world. The origin of the name is as obscure as the department that runs it. In itself, the word 'entity' has no connotations, good or bad. An entity list is merely a list of things, but in stark reality, it is a blacklist. Set up during the Cold War, until the Trump era it was little used. It is now extremely busy, devoting most of its energies to China and Russia.

The Entity List was directed at Huawei's business as a whole. All the indications suggest that the Trump administration's intention was not just to crimp Huawei but to cripple it. In practice, the ban turned out to be a slow burn. Google's immediate compliance with the Entity List – the removal of Google Mobile Services beyond current models – was a serious blow to Huawei's overseas sales. But it had no impact in China, the company's largest market. The Commerce Department also granted Huawei several consecutive 90-day reprieves for components. Consequently, the Entity List restrictions did not become permanent until 1 April 2020, by which time Huawei had spent nearly a year living without the full ramifications of the ban. The Department of Commerce used this time to devise a licensing system, whereby US firms could obtain approval for the sale of components and software to Huawei that posed no security threat to the United

States. Numerous applications for licences were received by the BIS, but none were granted.

For Huawei – and for CML – the axe came down decisively in May 2020. The US authorities announced a further measure that dealt a body blow to its smartphone business. As it worked through the operational mechanics of the Entity List, the Commerce Department came to appreciate just how much of a hold American software, design expertise and equipment technology had over the manufacture of leading-edge semiconductors. This EDA leverage could be employed to curtail Huawei's chip design ambitions. The Trump administration accordingly made Huawei the subject of a formerly arcane provision of an export control law called the Foreign Direct Product Rule. Under the FDPR, foreign-made items are subject to American controls wherever they are produced if they use American technology or software. Virtually all equipment in this sector uses American technology somewhere. These export rules – aimed at cutting the ground from under Chinese advanced semiconductor technology – were updated and consolidated by the BIS in October 2022.

Huawei's range of phones, including its sub-brands, Honor and Nova, overwhelmingly used Kirin SoCs, designed in-house by HiSilicon and fabricated under contract by TSMC. The latest generation of SoCs, the Kirin 9000 destined for the Mate 40, used a 5nm process, comparable to the level of integration offered by Apple, Samsung and Qualcomm. The application of the FDPR extended the controls to semiconductors made anywhere in the world with American technology, vastly expanding the reach of the restrictions. Effectively, this meant all advanced semiconductors. TSMC had no alternative but to comply. Huawei was consequently denied access to the fabrication of the SoC that powered its smartphones, especially the flagship P and Mate series. Domestic substitution was impossible. SMIC, the leading Chinese semiconductor foundry, was simply incapable of manufacturing chips below a 14nm process node. In one bound, Huawei had been transformed from a *firma non grata* into a *bête noire.*

Huawei had no choice but to acknowledge publicly that, from 15 September 2020, it would no longer be able to source Kirin processors from TSMC. Phone production accordingly plummeted in the final quarter. To relieve pressure on its handset business, in November 2020, Huawei sold its predominately online, mid-market Honor brand to a consortium for about $15bn. Honor's independent ownership would permit it to sell handsets with Google Mobile Services pre-loaded and buy SoCs to power them from Qualcomm. Huawei retained its Nova brand, mid-range smartphones targeted at the younger generation via celebrity endorsements and featuring superior design and excellent graphics. Honor, the cast-off child, is now a significant smartphone manufacturer in its own right, selling in China and internationally. Its success has led to talk of an IPO.

Shenzhen Zhixin New Information Technology, Honor's new parent company, was co-founded by Shenzhen Smart City Technology Development Group and over 30 agents and dealers of the Honor brand. In practice, the overwhelming majority of the shares are held by the Development Group, a management company owned by Shenzhen's district government. In short, Honor is state controlled. At the time of the sale, Honor was responsible for perhaps one-third of Huawei's handset output. Honor has since received approval to acquire the chips it needs and to access Google software for its handsets.

By the end of 2020 the US authorities had relented, and the restrictions had been eased, allowing Huawei to purchase from Qualcomm. Prior to this, Huawei and Qualcomm announced that they had reached a settlement in a two-year-old patent dispute, resulting in a $1.8bn payment by Huawei. It thus became possible for a US supplier to Huawei to obtain approval for certain items. Qualcomm was granted a government licence to sell Snapdragon SoCs, with the significant proviso that these chips did not include 5G capability – significant because the coming generation of premium phones featured 5G. Huawei could resume making handsets but with one hand tied behind its back. The company had no alternative but to progressively replace the Kirin SoCs in its handsets with Qualcomm's 4G-enabled Snapdragon chipsets.

The US ban on components that could be sold to Huawei was not a one-way street. Suppliers of these components were naturally unhappy at the prospect of losing business and made their views known to the federal government. Chinese handset producers were – and are – important customers for US-designed/made parts. Qualcomm especially, as we saw in Chapter 7, is heavily reliant on China for sales.

2020: PANDEMIC AND PRESSURE

Handset production at Huawei initially held up well. Until May 2020 – when the FDPR bombshell hit – the company felt that it could make 200mn phones in the year, not that far short of its 2019 production figure. Anticipating that the USA would take measures to restrict its activities, Huawei had stocked up with components and reduced its purchases of US-made parts. Covid-19 affected activity in the world at large but not so much in China, where factory activity continued amid highly restrictive containment measures.

In the marketplace, Huawei forged ahead and, according to the consultancy firms Canalys and Counterpoint, actually sold more handsets globally than any other vendor in the second quarter of 2020, just pipping Samsung at the post. While not denying the scale of Huawei's achievement, in truth, the company benefited from a happy conjunction of circumstances. The pandemic reduced worldwide smartphone sales by 20 per cent. But China held up comparatively well, declining by only 7 per cent. With its commanding position in the Chinese market – a more than 40 per cent share – Huawei shipped a large number of handsets. By contrast, Samsung was particularly hard hit. It derived no benefit from the relative buoyancy of the Chinese market because of the company's historically weak position there. However, by the third quarter, Samsung had regained the top spot internationally. The final quarter of 2020 reflected the full impact of sanctions. The 183mn handsets Huawei produced in the year, compared with 240mn in the previous year, did not look so bad but it masked a rapid deterioration in its position.

How did CML fare in all this? By the end of the first quarter of 2020, the impact of sanctions and Covid was clear. Volumes were down and CML's licensees, TDK and Billu, were under pressure to cut prices. This was despite an excellent reception for the Huawei P40 Pro, launched in March, which had reached the top of the DXOMARK league table. Shipments of actuators in the first quarter were 60 per cent down on budget. The CML 2020 projection was downgraded from a turnover of £12mn – accompanied by a small profit – to revenues of £7mn and a loss. For a while, there seemed to be some hope that the volume shortfall could be made up later in 2020, but any such expectation was dashed by the news in May that Huawei would be prevented from accessing its own Kirin chips. By mid-year, CML had taken measures to cut costs, including a hiring freeze.

Meanwhile the licence agreement signed in May 2019 with Huawei for the low-cost AF+OIS 8-wire camera actuator, utilising its preferred supply chain of Alps, TDK and Billu, was moving ahead. CML was looking forward to Huawei incorporating it into the forthcoming Mate 40, as a trial model before putting SMA into the rest of the range. Management had negotiated a higher per unit royalty rate, so revenues would be commensurately higher than for the current 4-wire product.

In the September 2019 Shareholder Update presentation, investors were told that the 8-wire project was progressing. Then …

everything went quiet. It was as if the project had stalled. The Mate 40 duly appeared on time, in October, powered by a Kirin SoC – drawn down from stock – and included the 8-wire in two premium models in the range. Shareholders then learnt in January 2021 that CML was still negotiating with Huawei. What happened during those months is not clear but must relate to the turmoil that consumed Huawei's smartphone business as it confronted the loss of the Kirin SoC. For CML, there was now no prospect of revenue recovery in the second half of the year.

What of the other projects CML was working on? As listed below, these covered a variety of promising areas. Underlying these initiatives, of course, was the pressing need for CML to reduce its dependence on Huawei.

- SMA 4-wire OIS actuators were beginning to be taken up by other smartphone manufacturers. The first of these design wins was from the Taiwanese maker, Asus, for the Zenfone 7 handset. It was followed by one from the Chinese phone maker, Meizu. Such design wins were nice to have but the volumes were small.
- LG Innotek, Apple's preferred camera supplier, expressed interest in a new, slimline AF actuator that CML had developed, capable of halving the height of the ever-expanding camera bump in current phones.
- Conventionally, OIS is employed to shift the lens in a camera module. An alternative method of achieving the same objective is moving the image sensor. This is known as sensor shift. It has some advantages. In May 2020 CML was approached by GalaxyCore, a Chinese image sensor company mentioned previously. GalaxyCore's engineers had the bright idea of combining an SMA OIS actuator with a new, larger image sensor they intended to develop. The resulting product would be highly cost competitive.

» In addition to lens shift and sensor shift there exists a third stabilisation technology: module tilt. As an option, it is most attractive as it offers five times the stabilisation of lens or sensor shift. Module tilt was starting to gain traction as an idea because, with the advent of 5G-enabled phones, users were shooting more video.
» CML continued to pursue haptics and 3D sensing. CML's haptics technology was shown to several major companies. Unfortunately, it proved to be too expensive and CML decided to pause its endeavours while the team re-evaluated its offering. By way of contrast, much excitement continued to surround 3D sensing and what it could do, which is the subject of the next section.

The chairman's review of the year in an annual report does not normally begin with an admission. CML's did.

> 'In preparing this statement for the annual report for 2020, I re-read what I wrote in last year's annual report. Its tone was upbeat and optimistic. The last paragraph read: "2019 was an enjoyable and successful year. I look forward to 2020 being even more so." Well 2020 wasn't enjoyable; and it wasn't uniformly successful for Cambridge Mechatronics Limited (CML).'

In fairness, at the beginning of 2020, the imprint of the May 2019 embargo was visible. What could not have been foreseen was the effect that the imposition of the FDPR rule in May 2020 would have on the supply of Kirin chips for the flagship models that employed SMA actuators in their camera systems.

AUGMENTED REALITY REVISITED

Within the smartphone industry, considerable excitement had been building up around Time-of-Flight rear cameras as a gateway to augmented reality (AR). What was CML's interest in all this? Encouraged by its involvement with Huawei on the Mate 30 Pro, the fertile research brains at CML had devised a way of increasing the range and resolution of the ToF sensor to create an even better augmented reality experience. As this involved combining optical components, actuators and software, CML entered into a three-way collaboration with Sony and Sunny Optical (in its module integrator role). It also developed a relationship with the image sensor division of Samsung. Clearly, this was a serious proposal. If successful, it could boost the capability of Android ToF sensors in the battle with Apple's LiDAR. In September 2020 CML told shareholders it was delivering samples to major smartphone brands of 3D-enhancement units – as the company liked to call them – offering an increased ToF operating range of more than 10 metres (compared with LiDAR's maximum range of 5 metres).

Unfortunately, the initial burst of optimism for 3D sensing and AR in handsets had cooled by 2021. Certain Android manufacturers, notably Samsung and Huawei, had dipped their toe in the water by installing ToF sensors on the rear of their flagship models but there was no groundswell of enthusiasm. Huawei might have continued to fly the ToF flag for Android but the US trade embargo threw everything up in the air. Samsung then dropped ToF from the back of the Galaxy S22, citing lack of customer appeal. There *has* been growth in the mobile usage of AR apps but certainly not at the level predicted. The consensus front 3D camera forecast remained much the same but, by 2021, uptake projections for 3D rear cameras had been massively downgraded.

The main factors behind the low take-up of AR, in retail especially, appear to be:

1. Lack of pull from the consumer. Overall, the experience available to the user via apps was not good enough. Meanwhile, companies veered towards opportunities for 3D sensing in non-mobile markets, such as smart door locks, robot vacuum cleaners and – the one with the largest potential – automotive.
2. LiDAR has remained expensive. The high cost has discouraged EV manufacturers from using it in automotive cameras. Although plain ToF was cheaper to install in a phone than LiDAR, it was still costly so could only be justified for top-of-the-range models.
3. Both Apple and Google continued to advance their proprietary software-based methods of facilitating AR on mobiles without ToF.

It is undeniable that a depth-camera-equipped phone improves AR. Response is immediate, and a ToF/LiDAR sensor is better at measuring distances and recognising surfaces. But the plain fact of the matter was that AR apps can be created for, and work adequately on, phones without a ToF sensing capability. Constant improvements in ARKit and ARCore were undercutting the argument that AR needed 3D-enabled handsets to realise its potential. Consequently, enthusiasm for ToF sensing in the Android camp waned. Whether Android vendors generally would decide to commit to ToF sensing – and CML's 3D-enhancement solution – remained very much up in the air.

23.

Vicissitudes and resurrection

SANCTIONS BITE

During Covid-19, enforced working – and playing – from home gave a big boost to fixed consumer electronic equipment, notably personal computers and televisions. Global smartphone shipments declined, having been on a gentle downward trend since 2017, but picked up a little in 2021. CML had a number of irons in the fire but the overarching issue remained the volume of phones – and therefore actuators – shipped by Huawei. In 2020 Huawei delivered nearly 190mn devices. Deducting, say, 60mn Honor handsets removed by the divestment implied a target of around 130mn devices to stay at the same output level. Huawei's own Kirin SoCs were no longer available – other than those in its dwindling stock – so the company became dependent on deliveries from Qualcomm of the Snapdragon equivalent, tweaked to ensure it could not support 5G.

In his review, CML's chairman talked of 2020 as a 'tough year', referring briefly and without elaboration, to the 'continued tribulations of our largest customer'. Quite how devastating the measures taken by the USA during 2019 and 2020 were for Huawei is laid out below. Apple, as the stand-out competitor at the upper end of the Chinese market, was the principal beneficiary.

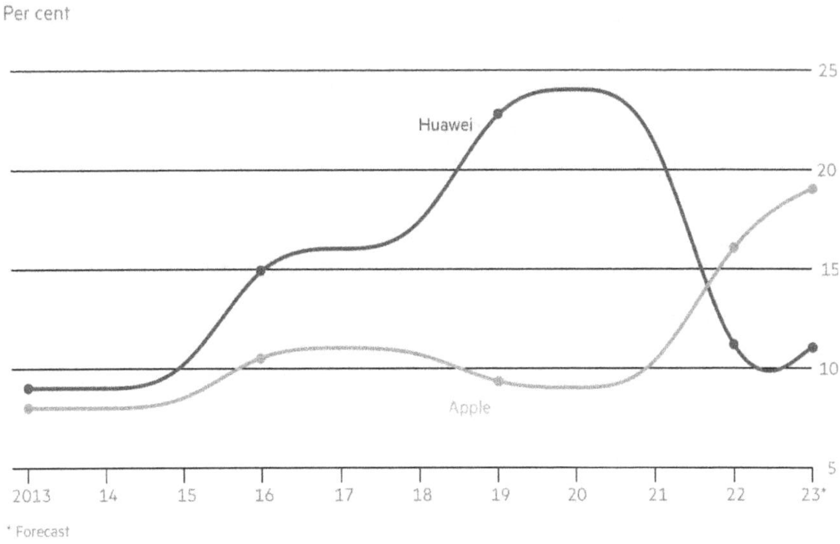

Source: Counterpoint Research

Following some technical problems, the Mate 40 was released in October 2020. The phone itself was once again well received, a typical comment from the techno-commentariat being:

> 'In a world where the HUAWEI Mate 40 Pro had Google access and widespread availability, it would have been a contender for the best smartphone of 2020.' ('HUAWEI Mate 40 Pro review: The full hardware package', Ryan-Thomas Shaw, Android Authority, 10 September 2022)

In these trying circumstances, Huawei chose to depart from its normal launch schedule. The P50 series was delayed from March to the end of July 2021. Some handsets used Snapdragon chips, some Kirin from inventory.

GOOD TIDINGS

The scale of the damage inflicted on CML's business by Huawei's troubles is summed up by the numbers in the table below. Production effectively ceased for two years.

	2019	2020	2021	2022
Actuators produced	24mn	No figure given	No figure given	6mn

Source: CML annual reports

As 2023 dawned, CML was, thankfully, greeted by some positive news. Alps Alpine emerged from the shadows to announce that it had commenced mass production of SMA actuators, supplementing its existing VCM actuator line-up. CML was not mentioned in the press release but everyone in the industry was aware of the source of Alps' new SMA technology. Better still, SMA received public endorsements from two highly regarded technology consulting firms: Counterpoint Research and Yole Group. Both praised SMA. Yole went furthest, stating unequivocally on its website that:

> 'SMA actuation stands out among existing actuation solutions.' ('Alps Alpine to shake up the camera module actuator market with SMA technology mass production', Florian Domengie, Peter Bonanno, Yole Group, 21 March 2023)

Against this favourable background, Alps mounted a promotional campaign to persuade smartphone manufacturers to adopt 8-wire SMA, with Apple front and centre of this effort. Unfortunately for Alps and CML, it transpired that Apple – having lagged behind the competition in telephoto cameras – was in the throes of evaluating actuator options for the periscope camera it planned to introduce on the iPhone 15. Replacing the main camera actuator was consequently low on the priority list. Zoom/periscope OIS cameras demand a different set of actuator characteristics compared with a primary camera, specifically a greater length of travel and a high force capability. SMA was not a natural candidate for this function as it did not have the necessary length of travel.

Apple's periscope camera emerged in 2023 in the iPhone 15 Pro Max. It took the form of a TetraPrism, Apple's own take on the periscope concept. (Note that this was four years after Huawei had introduced periscope technology and three years after Samsung.) Responsibility for Apple's tardiness can be laid at the door of the OIS actuator. For reasons mentioned previously, OIS is essential for zoom cameras. Alps Alpine and MinebeaMitsumi, Apple's established OIS suppliers, could not provide the required actuator. Accordingly, for its TetraPrism module, Apple turned to a refinement of VCM technology developed by Samsung called a ball guide actuator (referred to previously, and explained, in Chapter 15). Made by the Samsung affiliate, SEMCO, and a Korean contractor, Jahwa Electronics, it had the advantage of familiarity, being manufactured by an existing supplier with an established, low-cost production base.

Apart from Apple's reluctance to become any more dependent on its major competitor, going down the ball guide route involved wending a way through a convoluted Samsung-controlled patent situation. Once accomplished, Apple turned to its long-term partner, LG Innotek, for the supply of ball guide actuators for its periscope camera module. To provide a second source, it did a deal with Jahwa Electronics despite the contractor being firmly within the Samsung orbit. Samsung, having devoted considerable resources to developing the ball guide alternative to conventional VCM actuators, determined to apply the technology to the primary camera module across its model range. It also announced its intention to sell ball guide actuators to the Chinese smartphone producers. This constituted a serious challenge to CML in its ambition to penetrate the Chinese primary camera market with SMA.

Price was a major issue here – VCM OIS undercut SMA. Ball guide actuators could be made cheaply on existing production lines. Conversations with the Chinese manufacturers reinforced the message on price. Uniformly, the response to CML's marketing overtures was that the companies were sold on SMA as a technology but were only prepared to pay a small premium over the price of the equivalent VCM. In response, CML urged its manufacturing licensees (Alps, TDK, Billu and now ZET) to take measures to bring costs down.

Fortunately for CML, in March 2023, Huawei came to a different conclusion. Having had ample opportunity to observe SMA operating in the field for five years, Huawei resolved to extend the use of SMA 8-wire AF+OIS actuators to its entire model range. This was, of course, excellent news but – due to sanctions – the potential number of units that Huawei could offer was only a fraction of the volume that a design win from Samsung (or Apple) could steer in CML's direction. Encouragingly, as well as Huawei, Honor was beginning to show interest in SMA for some of its future models. In anticipation of this increased level of demand, CML's four manufacturing licensees planned to create a combined capacity of 40mn units a year by the end of 2023.

HUAWEI PULLS AN AUGUST SURPRISE

By the autumn of 2022 Huawei had recovered sufficiently – with Qualcomm as its SoC supplier – to resume its normal annual launch schedule for the Chinese market (to which it had effectively been confined). The Snapdragon-powered Mate 50 and P 60 series appeared on time, featuring Huawei's own HarmonyOS operating system rather than Google's GMS.

Then suddenly, without any of the usual razzmatazz that accompanied a Huawei flagship launch, the Mate 60 was slipped out at the end of August 2023 in the form of an announcement on the Huawei website. What excited the army of dedicated Huawei fans was that the phone appeared to have capabilities beyond those dictated by American sanctions. News of the new phone emerged during a week in which Gina Raimondo, the US Commerce Secretary, was in Beijing for talks with her Chinese counterpart. Was the timing a coincidence? Almost certainly not.

Huawei's mysterious new phone gave rise to a frenzy of speculation. Its supposed performance characteristics quickly became part of the national conversation. Did it have 5G capability as rumoured (specifically prohibited under the US embargo)? Early reviews from China suggested that this device offered 5G upload and download speeds. And what SoC was it using? Clearly not Snapdragon but a chip of its own making. Huawei remained uncommunicative. It was hoped that all would be revealed on 25 September at a glitzy event in a stadium showcasing a series of new products from the company and watched by millions online. To the frustration of observers – and the US administration – nothing of significance was disclosed.

Instead of the usual multi-week delay between announcement and the commencement of deliveries, the phone was fully on sale within four days. Enthusiasm among the Chinese public for this new handset and its purported capabilities was palpable. Citizens turned out in droves to support the national champion. People queued outside Huawei stores across China. Online availability soon ran out. Within weeks, an analysis by the Canadian technology consultancy, TechInsights, found that the 5G-capable Kirin 9000s SoC inside Huawei's Mate 60 was manufactured by China's leading foundry company, Semiconductor

Manufacturing International Corporation (SMIC). It too was on the Entity List, having been added in 2020. Huawei's handsets could now match the 5G capability that Apple, Xiaomi, Oppo and others were promoting to domestic consumers. The bottom line was that Huawei had been resurrected as a smartphone competitor.

The American embargo was deliberately designed to prevent this happening, by curbing Chinese access to critical chipmaking equipment and tools. Yet Huawei had somehow circumvented these restrictions. Cue much consternation and soul-searching in Washington. How had the trick been accomplished? Huawei remained tight-lipped about the characteristics of the 9000s and how it had been manufactured by SMIC but gradually more information dribbled out.

The original HiSilicon-designed 9000 SoC in 2020 had been fabricated (by TSMC) on a 5nm process. For the 9000s, SMIC had used a less advanced 7nm process node. Until then, SMIC had been known for fabricating 14nm node chips at best. In essence, SMIC, in close collaboration with Huawei, had found a way of producing advanced chips with outdated equipment.

As explained in Chapter 3, the ultimate chokepoint for anyone seeking to manufacture cutting-edge semiconductors is lithography – using light to print tiny patterns on silicon. The Dutch company ASML is uniquely capable of constructing the incredibly sophisticated machines to do this. The current technology, used by TSMC and Samsung, is extreme ultraviolet (EUV), which superseded deep ultraviolet (DUV). Under the US embargo, ASML agreed not to supply EUV machines to any Chinese entity, only DUV equipment. But SMIC found a way to use DUV to achieve a 7nm outcome. To achieve this, HiSilicon had to make numerous design compromises. The upshot is that the process involved is more cumbersome, and yields are lower. The chip's performance was also affected. (One issue was that, in the jargon, it 'ran hot' and the Mate 60 handset was fitted with an extra-large vapour chamber cooling system to improve heat dissipation.)

Nevertheless, by dint of engineering ingenuity, hard work and perseverance, working together, SMIC and Huawei had been able to create an SoC that broadly matched the performance of the equivalent Snapdragon three years earlier. The Mate 60 SoC was not on a par with offerings from Apple, Qualcomm and MediaTek but it was certainly good enough to permit Huawei to get back in the game. For its part, SMIC appeared set firmly on the path of upgrading its process technology from the current 7nm process node to 5nm and below, so as to be in a position to supply SoCs for Huawei's future flagship needs.

As for the restrictions that were intended to prevent Huawei reaching this point, the American administration was forced to confront the fact that they were porous. Chinese foundries had stockpiled ahead of the ban. SMIC managed to scrape together enough equipment from existing plants and those obtained before sanctions. Necessary upstream inputs other than EUV machines had been secured via third countries or third-party corporate intermediaries. There were also indications that key Western and Japanese chipmaking machine and material suppliers – for whom China is an important market – had been less than assiduous in monitoring the ultimate destination of their products. Recognising the weakness in the controls, in August 2024, the US authorities persuaded the Dutch government to restrict the sale of ASML's more sophisticated DUV lithography machines to China, invoking the FDPR rule relating to US-made parts.

As well as a focus on cutting-edge semiconductors, China is equally intent on expanding its mass production capabilities for mature node (above 30nm) chips. The country is the pre-eminent global importer of semiconductors because it is home to a large proportion of the world's electronics manufacturing capability. This strategy has a specific aim of ensuring a home-grown chip supply feed to its burgeoning EV (electric vehicle) industry. Apart from SMIC, at least a dozen smaller chipmakers have been busily buying permitted semiconductor production equipment. According to the industry group SEMI (an international association comprising companies involved in the electronics design and manufacturing supply chain), Chinese firms accounted for a third of all chipmaking equipment purchases globally in 2025.

There is a second part to the equation that results in a successful smartphone SoC. Creating a design that works is one thing. Producing it in sufficient quantities to satisfy the voracious appetite of an in-demand smartphone is quite another. The key here is yield. Once a design is bedded down, an established wafer fab will have final yields in the region of 80–90 per cent. The initial yield that SMIC was able to obtain was the subject of much speculation. Industry experts estimated it at 50 per cent and possibly less than 30 per cent, as per the following comment.

> 'The move to create more advanced chips has incurred additional costs, however. Three people close to Chinese chip companies said that SMIC was having to charge 40 to 50 per cent more for products from its 5nm and 7nm fabrication nodes than Taiwan's TSMC does at the same nodes. ('China on cusp of next-generation chip production despite US curbs', *Financial Times*, 6 February 2024)

To ensure that Huawei did not pay a prohibitive price for its smartphone chips, the Chinese government provided substantial subsidies.

As Chinese consumers clamoured to buy the 9000s-powered Mate 60, initial SoC production could not meet demand. These issues were soon resolved, suggesting that yields had improved. Huawei had a good final quarter of 2023, beating Apple in the domestic marketplace, aided by a government prohibition on iPhone purchases by employees of state-run enterprises. In the event, Huawei sold 14mn Mate 60s over its lifetime.

Apple remains a vibrant contender in the Chinese market, but clouds are gathering. Despite Apple having forged a partnership with Alibaba in early 2025, the regulator has been slow to approve the local rollout of AI features on the iPhone due to China's strict rules on foreign AI tools. Having held the lion's share of the Chinese high-end market, Apple's position came under pressure, primarily to the benefit of Huawei. Then, in late 2025, the launch of the iPhone 17 range revived Apple's fortunes, despite the lack of AI features offered by local rivals.

How much progress has Huawei made in securing supplies of a globally competitive SoC to power its handsets? Huawei's next phone, the Pura 70 (as opposed to just 'P') with a HiSilicon-designed and SMIC-manufactured SoC, the Kirin 9010, appeared on schedule in April 2024. The Mate 70, however, did not appear until November 2024, two months later than the normal September launch date. No explanation was given. There was widespread speculation that the delay reflected yield problems with the latest version of the Kirin 9000 SoC, such that SMIC needed more time to produce the chip in sufficient quantities before launch. Perhaps reflecting this, Huawei's own figures suggest it sold fewer Mate 70s than Mate 60s.

ONE DOWN, ONE UP

The last chapter mentioned LG Innotek's interest in CML's ultra-slim autofocus actuator, designed to reduce the bump in conventional handsets and offer a camera solution for folding handsets. LG Innotek was, after all, the principal camera module supplier to the iPhone. After the failure to secure a position with Samsung and the enforced contraction in Huawei's offtake, an opportunity to get closer to LG Innotek was a ray of light. The relationship turned into something tangible in September 2020 with an agreement for an evaluation licence, allowing LG Innotek to build and test SMA actuators. The hope was that this would lead to a 'platform' licence granting LG Innotek broad access to CML's SMA technology. However, the overarching agreement CML was seeking remained elusive. Negotiations dragged on into 2022. It then became apparent that LG Innotek had lost interest.

We can only speculate on why the relationship never amounted to much. It must be germane that LG Innotek was going through a boom period. In 2021, sales grew by 56 per cent, with improved profitability to match. The following year sales increased by a further 31 per cent. Every quarterly report contained a reference to the 'supply of new flagship models to overseas customers'. Use of the plural was, of course, disingenuous. There was only one overseas customer: the one that shall not be named. Most probably, the desire to access a new actuator technology quietly slipped off the priority list.

In contrast to the slow and ultimately futile courtship of LG Innotek, the licence relationship with Alps Alpine, initiated in 2019 at the behest of Huawei, developed into a mutually beneficial connection. Alps moved fast, impressing the team from CML. By the second half of 2020 it was capable of making 8-wire SMA actuators in volume. In short order, Alps Alpine cemented the relationship by buying a small equity stake in CML (details in the next chapter). From CML's perspective, the association with Alps had the valuable secondary benefit that, as the leading iPhone VCM OIS supplier, it represented a potential route into Apple.

To ensure a second source, Huawei had insisted that Alps transfer its technology to the Chinese VCM actuator maker, ZET (Liaoning Zhonglan Electronic Technology Co., to give it its full name). ZET is minority owned by Huawei, an existing customer. All this took time. Coupled with the delays imposed by sanctions, it was not until late

2021 that the 8-wire began to be inserted into the Mate 40. To get this far had taken the best part of 18 months. And it meant that the 8-wire had missed the boat for the Huawei P50.

Despite Huawei's disappointingly slow take-up, the overall relationship between Alps Alpine and CML continued to develop fruitfully. Alps perceived that embracing SMA would give the company another string to its bow in its effort to defend and grow its all-important smartphone actuator business at a time of technological flux. In September 2022, a high-level deputation from Alps Alpine attended a board meeting of CML in Cambridge for a wide-ranging discussion on what the two companies were doing together and could do together in the future.

Overall, the company was moving forward. A snapshot at September 2021 showed that CML licensees were producing actuators for 19 handsets and had achieved design wins at all the major Chinese smartphone vendors. But individual model volumes were small. For the year as a whole, royalty income fell by 30 per cent.

In the last chapter we described several projects that CML was working on. What progress was being made on these?

SENSOR SHIFT

Why did smartphone manufacturers start to express interest in sensor shift stabilisation? After all, lens unit stabilisation was well established and worked. In this, Apple set the trend, steering away from conventional lens stabilisation in the iPhone 13 range, launched in September 2021. One reason was the trend towards heavier lenses. Image sensors are relatively light and therefore easier to move. Adjusting directly where the image capture actually happens – at the sensor – produces a superior result. A more important advantage was the ability to stabilise in an additional dimension. Camera movement can be split into linear movement (sideways, up/down) and rotational movement around the optical axis. Lens shift OIS cannot compensate for rotational movement, but sensor shift OIS can. This is known as '5-axis stabilisation'. Rotation is the fifth axis.

Once again, following the pattern set by FaceID and LiDAR, the Apple sensor shift solution was complicated, space-consuming and expensive. Apple's introduction of sensor shift did not come as a surprise to the Android ecosystem or to CML. It was, after all, an established digital camera feature that smartphones would be bound to try and replicate. CML had prepared for its appearance in the iPhone 13 and was ready with a simpler and lower cost Android solution.

The first handset incorporating CML's proprietary sensor shift technology was the Oppo Find X5 flagship, released in February 2022. Oppo's promotional literature for this device contained a tangential reference to CML by describing the image sensor as 'driven by shape memory alloys'. Subsequently, a few more flagship phones substituted sensor shift for lens shift in their primary cameras, but the adoption of sensor shift by the Android community was gradual and patchy. Huawei did not incorporate sensor shift until the P60, in 2023. Samsung, characteristically, busied itself exploring sensor shift with a view to developing its own solution.

The licence arrangement signed in mid-2020 with the Chinese image sensor company GalaxyCore, giving it access to CML's sensor shift OIS technology, got off to a good start. The objective was to create a high-end package that could be sold for a considerably lower price than the current offerings from Sony and Samsung. GalaxyCore engineers believed that combining a 4-wire SMA OIS actuator with a to-be-developed 13MP image sensor would enable the creation of a high-resolution sensor, with a performance equivalent to 52MP. Production was scheduled for 2022. A figure of 20mn units was pencilled in for that year. Despite its extensive experience of making low-cost image sensors, GalaxyCore hit technical problems with the new, larger image sensor it was developing. The timetable slipped … and then slipped again. The good news was that Xiaomi had been signed up as the launch customer. By September it looked as if the project would reach the production phase a year later than envisaged.

MODULE TILT

The next stage in the onward march of image stabilisation technology was briefly described in the last chapter. In March 2021 CML reported to shareholders that module tilt OIS was 'in development'. Module tilt is so-called because, instead of moving the lens or the image sensor to compensate for shake, the entire camera module is stabilised. Why the need to introduce yet a further type of optical stabilisation beyond lens shift and sensor shift? The short answer is: video.

Increasingly, vloggers and videographers wanted to be able to record smooth videos on their smartphone while walking or running around, confident that the camera will stabilise their footage whatever they are doing. Long videos are prone to camera shake, especially when using a zoom lens. Among professionals, and some amateurs, the standard accessory for this purpose is a gimbal, a mechanical stabiliser that, through a combination of sensors and electric motors, keeps the phone steady regardless of how you move your hand, arm or body. However, a gimbal represents a significant additional cost.

Matching what a gimbal can do for smartphone videography is the holy grail of the stabilisation business. The team at CML came up with a gimbal-equivalent module-tilt solution based on the 8-wire SMA actuator. By the end of 2022 CML was confident it had developed a model with exceptional performance, completely eliminating video blur. It was promoted to Huawei. It was also marketed more generally, including to Samsung.

3D SENSING

For CML, the original assumption on 3D was a simple one: that a multiplicity of appealing augmented reality apps would drive demand for the inclusion of ToF rear sensors on Android phones. But depth sensing can be accomplished by clever use of a phone's camera and sensors. Adding a ToF sensor improves the AR experience, but it is not essential. Confronted with this hard truth, management reluctantly concluded that there would be no significant market on Android phones for the 3D-enhancement ToF technology it had to offer.

24.

The brave new world of extended reality

VIRTUAL AND AUGMENTED REALITY: WEARABLES

When, in mid-2021, it became clear that rear ToF/3D cameras on smartphones were not going to be the runaway success CML had hoped for, the company shifted direction. Not in corporate terms but in regard to how SMA might be employed in the context of augmented reality. As noted previously, CML had always considered its heavily patented technology to be a platform, granting SMA the ability to pivot towards application areas beyond smartphones. Around 2020, handset designers began to imagine what the phone of the future might look like. Would it actually be a phone or some other form factor? Form factor covers a range of formats, from folding phones through glasses or headsets to smart watches and wristbands, often worn rather than held.

Augmented reality had not gone away. It was common ground in the tech sphere that AR, with its ability to expand the consumer experience through information and entertainment add-on features, had great potential to enhance user engagement (and drive device sales). VR and AR are often written about, or spoken of, in the same breath (as in 'VR/AR'). There is indeed a strong sibling relationship between them. VR is the slightly older sibling, with (metaphorically) an introverted personality as VR positions the user in an entirely internal world.

VR has found application in the consumer space, predominately for gaming. Gaming is a niche activity – albeit a large niche – but it has proved hard to expand beyond the cohort of enthusiastic early adopters. Most of the running in consumer VR has been made by Meta with its Quest range of headsets, aided by its library of apps and games. Both virtual and augmented reality rely on digitally created images, sounds and texts for information and entertainment. The difference between them is that AR, the younger sibling, is possessed of a deep desire to see what is happening in the surrounding world at the same time as viewing computer-generated information.

One of the principal factors discouraging consumers from using AR on smartphones for, especially, online shopping was – and still is – the need to hold the phone for an extended period of time. A headset or, better still, lightweight 'smart' glasses able to send and receive data via the internet would be preferable, allowing both hands to be free. Ideas about the feasibility of wearable technology – as this category is now called – have been around since the mid-2010s. By the end of the decade, items like smart watches and fitness trackers had become commonplace. Since then, wearable computing has become a fast-expanding sector within consumer electronics. Wearables covers products that can be worn as accessories, embedded in clothing, implanted in the user's body or even tattooed on the skin.

Google's early foray into the wearables field is a cautionary tale. In 2013 the company launched Google Glass. For $1,500, the spectacles-like product offered a display and a built-in camera, allowing users to see information and capture images of their environment. Google Glass was not an AR device but a step on the road towards one. Rather than promoting it as a prototype, the marketing campaign and the high price positioned it as a fully functioning upmarket item, which it clearly was not. Privacy issues – the ability to take photos without the subject's knowledge or permission – dogged the product. Google Glass became the subject of public derision (as in 'Don't be a Glasshole') and was withdrawn a year after launch.

Undeterred, in 2017, Google reintroduced Google Glass for enterprise (business) use. It too failed to gain traction and sales ceased in March 2023. Indeed, Google has acquired a reputation for U-turns, which may be due to the diffuse manner in which the company operates, with multiple, overlapping decision points. A group of Google-watchers have gone so far as to construct a tongue-in-cheek website entitled Killed by Google – a comprehensive list of discontinued Google products, services, devices and apps. It contains 300 items, including 25 hardware projects.

One company, in particular, epitomised the boundless optimism of the AR apostles: Magic Leap. This turned out to be a less than inspired name: Magic Stumble would have been more appropriate. At inception in 2018, Magic Leap was valued at $6.4bn, having obtained funding from several heavyweight investors, including Google and Qualcomm. The company's consumer headsets attracted much attention but few sales. In 2021, the Saudi Arabian Public Investment Fund bought a controlling stake and injected further much-needed cash. Like Google, Magic Leap repositioned itself as a supplier of AR headsets to the enterprise market. It has yet to prove itself. However, in 2023, it transpired that salvation – to a degree at least – was at hand from an unexpected quarter. Magic Leap had invested heavily in optics R&D and had created state-of-the-art production facilities in Florida and Mexico, generating numerous patents. The resulting IP and production capacity is in demand from competitor AR headset/glasses makers.

Since 2020, various trends and technologies in the wearable space have coalesced and acted as an incentive for tech companies to step up their AR headset/glasses efforts. Lower latency as a consequence of faster 5G network speeds boosts the immediacy of AR in operation. Software for

AR has gradually become more sophisticated and compelling. Advances in components permit developers to contemplate the ultimate dream of a wearable device that is lightweight, highly capable, with a long battery life and offered at a price point that opens up the mass market.

AR AND SMA

Early in 2021, CML perceived that its SMA technology could play a useful role in AR hardware and embarked on a marketing campaign to convince the major tech companies based on the US West Coast of this. By the end of the year these marketing efforts had borne fruit. There was an extremely positive response from Snap, the parent company of Snapchat. Although in essence a software-based, social media platform aimed at 18–34-year-olds, in 2017 it declared unequivocally in its IPO filing document that 'Snap Inc. is a camera company'.

Since then, Snap's trademarked Spectacles series of augmented reality smart glasses have gone through several iterations. They are essentially aimed at the 400mn daily active users of Snapchat. Spectacles were initially sold to the general public but, in order to obtain feedback, Snap then spent four years renting them to app developers. In June 2025, the company announced the launch of its renamed sixth generation AR glasses – Specs, projected to go on sale to the general public in 2026. Hardware details were scarce and, critically, there was no information on the selling price. Snap believes that the capabilities of the latest version of its dedicated operating system, Snap OS 2.0, and the variety of experiences for Specs that its army of 400,000 developers is building, will give the company an edge.

Snap's engineers could see that there were various points at which next generation Spectacles' requirements and CML's SMA technology, emphasising its autofocus capabilities, potentially intersected. Pre-eminent among these was the need for a small, low-power actuator to extend the limited battery life of the product. Further possible applications for SMA in future products included:

- Focus adjustment
- Varifocal displays
- Autofocus in cameras
- The athermalisation of AR devices (the need to maintain the correct lens focus in response to a change in the focal length due to temperature change)

A wide-ranging discussion ensued, culminating in a formal agreement at the end of the year to collaborate on a broad front. In the first quarter of 2022, CML started to perform engineering work under contract to Snap, for which it received stage payments.

CML's promotional trip to West Coast VR/AR firms also elicited a positive response from a second major tech company. By early 2022, there was an agreement with CML to do engineering work on smart glasses on aspects that could benefit from its advantages, especially low power consumption, and also on haptics. Though smaller and less wide-ranging than the Snap arrangement, it too produced upfront cash followed by staged payments. It is important to remember that engineering contracts, while nice to have, are not in themselves the basis of a viable business. The pay-off comes when – or rather if – these relationships lead to production volumes and a per unit royalty.

MR AND XR

The field that CML had entered via these R&D contracts was all very new. Further buzzwords began to appear in the posts of the techno-commentariat, specifically mixed reality (MR) and extended reality (XR). Firms with a strong base in VR, notably Meta, began to envisage a fusion of VR and AR. At the same time, numerous other companies with electronics expertise started to develop and sell so-called smart glasses. These offered the wearer some useful phone-like features but not necessarily an AR capability.

The terminology in this nascent field is still in a state of flux. AR particularly is ill-defined. Mixed reality (MR) blocks the user's vision but provides a view of the real world via outward-facing cameras. It blends the physical and digital worlds using both VR and AR, including hardware and graphics in a sort of hybrid reality. The technology allows users to interact with both the physical and digital worlds simultaneously, melding the two together. For example, in MR, a digital book can be anchored to a real-world location, like a table. When you want to read it, you put on your headset, pick up the 'book' with your real hand and turn the 'pages' with your real fingers.

There are therefore three ways in which a user can experience the contents of a headset or smart glasses: VR, AR and MR. The increasingly used umbrella term encompassing all these techniques is extended reality, or XR. Meta has moved away from 'pure' VR in the direction of AR with its latest generation of Quest headsets. The Mixed Reality Meta Quest 3, launched in September 2023 at a price of $500, looked like this:

Source: TechCrunch

The Meta website described the Quest 3 as a device that 'keeps your surroundings in sight, while stereoscopic depth perception allows you to interact with characters or objects in the room around you.'

Apple, unlike Meta, never went down the VR route. Instead, for several years, it has been working on an AR headset. Apple's Vision Pro augmented reality headset (below) was announced in June 2023. It uses 12 cameras and a multiplicity of sensors to make it feel like you are looking right through a screen, into the room you are in.

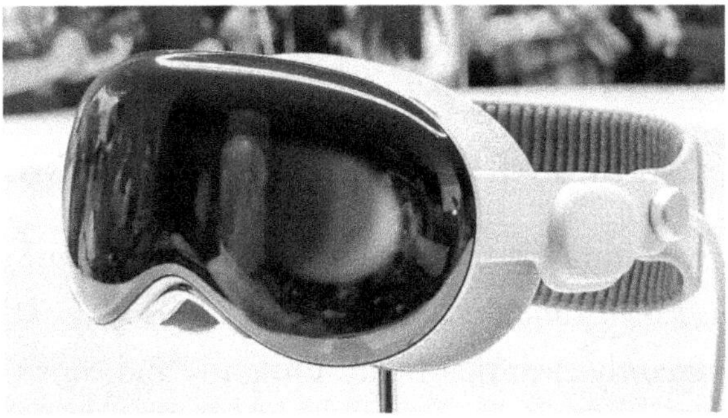

Source: Apple Vision Pro CNET, 7 June 2023.

Apple's guiding principle is that you are never isolated from the people around you, even when using the headset in VR mode to view something. Nor are they isolated from you. For example, Vision Pro's EyeSight feature lets a person nearby see the eyes of the headset wearer on the device's outer screen in an attempt to simulate eye contact. Conversely, the only way of attracting the attention of someone wearing a Meta Quest VR headset is signalling them to remove it. The Vision Pro is undeniably impressive but is complicated to manufacture and accordingly, at $3,500, expensive. Apple is steadfast in its aversion to calling the Vision Pro anything other than an AR headset. Others would refer to it as a mixed reality device. But VR, MR and XR are not part of the Apple lexicon. Instead, Apple prefers 'Spatial

Video'. Aimed primarily at developers, the Vision Pro has had a limited impact. It is too expensive, too bulky, uncomfortable to wear for any length of time and lacks appealing software.

On-screen AR applications can provide real-time information, for example, about tourist attractions, overlaying historical facts and user-generated reviews onto the user's environment. But the plain fact is that none of this is, as yet, immersive or compelling. Lack of suitable content is an issue but, at root, the problem is that appropriate and affordable hardware does not yet exist. Major constraints include on-device processing and battery life.

SMART GLASSES ARE THE FUTURE ... PROBABLY

Within the tech industry there is a general belief that the smartphone will be replaced in time by some form of smart glasses. Early on, it became abundantly clear that bulky, unsightly and constricting headsets were never going to be a mass-market consumer device. No one expects the transition to occur quickly. There are numerous technical hurdles to be overcome. CML is positioning itself to play a minor – but potentially significant – role in whatever emerges.

The widespread conviction that smart glasses will, in time, replace the 'slab' phone is stimulating interest among numerous companies. Consequently, it is an extremely crowded field. According to the Indian technology consulting firm Tracxn, which maintains a database of startups worldwide, at October 2023, there were 245 active participants in the smart glasses arena. Anxious not to be left behind, the smartphone SoC providers, Qualcomm and MediaTek, have directed resources towards creating the specific processing power in a tiny package that smart glasses require.

What are smart glasses? The term is a broad one, spanning a range of functionalities. Normally, to work, they need to be paired with a phone. In essence, they are wearable devices that can present information to the user. This could be a text message, the name of an incoming phone caller or directions, all of which is shown in 2D. At root, the term refers to computerised eyewear with an internet connection that offers basic handset features such as capturing photos, viewing notifications or making phone calls. More recent smart glasses also offer high-quality audio and AI assistant capabilities. Indeed, AI-powered glasses are gaining ground on basic smart glasses.

Some consumer smart glasses offer a head-up display (HUD), a small display that projects information in the user's line of sight, such as notifications, turn-by-turn directions, AI output or translations. Smart glasses are also pushing a feature that allows the user, when connected to a phone, tablet or laptop, to watch a film or video or play a game on a very large virtual screen that appears to be several metres in front of them. Many of the firms selling smart glasses sprinkle AR liberally over their marketing material, in an attempt to capture the aura surrounding AR in the public mind.

Augmented reality shows digital information as if it exists in reality. That means that smart glasses begin to become AR glasses when they are sufficiently adept at sensing your surroundings to present information in such a way that it feels like it is present, not merely projected onto a transparent screen. Mixed reality glasses offering complex 3D models and animations can create a convincing experience where, for example, virtual items hide behind real objects or sit on your furniture. MR glasses are expected to become the central XR platform over the next two decades, replacing AR glasses, which lack immersive or spatial capabilities.

HEADSETS AND GLASSES: PARALLEL PROJECTS

Meta is at the forefront of the development of VR and MR headsets. In parallel, it is doing much work on smart glasses. All of this is the responsibility of its subsidiary, Reality Labs, set up in August 2020 to advance Meta's XR technologies, both hardware and software. Funded by its core advertising business, Meta's commitment represents a huge, impressive and extremely expensive bet on the long-term future of XR. Notwithstanding some revenue contribution from the sale of Quest headsets and other hardware items, losses at Reality Labs have mounted, leading to a degree of shareholder unease. At the end of 2025, the cumulative deficit was close to $80bn. The technologies being developed are still at an early stage. From remarks made by Mark Zuckerberg, it is clear that the company does not expect meaningful profitability from its investment before 2030.

In smart glasses, Meta achieved a significant breakthrough in 2024 with its Ray-Ban Meta glasses. This product was the result of an imaginative move to combine Meta's technology with the design nous and strong brand recognition of the Franco–Italian firm, EssilorLuxottica, a force to be reckoned with in the spectacles world. In North America, its Ray-Ban Aviator brand is iconic. Ray-Ban Metas sold well. Priced at $299 at launch, Meta Ray-Ban spectacles are pictured below.

Source: 'Why Meta's Ray-Ban Smart Glasses are my favorite tech purchase this year', Prakhar Khanna/ZDNET, 13 May 2024

Ray-Ban Meta glasses are *not* AR glasses but a pair of aesthetically pleasing sunglasses that double up as a camera and a pair of headphones, allowing the wearer to make calls, record videos, livestream to Instagram and Facebook and send texts and images. The processing power is provided by a Qualcomm Snapdragon SoC specifically developed for use in smart glasses. In 2025 the appeal was considerably enhanced by the addition of AI, allowing users to interact with their surroundings by asking questions about what they can see and receiving answers. Ray-Ban Meta AI glasses also offer real-time speech translation between languages. In September 2025, Meta launched an upgraded product: Meta Ray-Ban Display glasses with a small display built into the right-hand lens. During 2025, Essilor-Luxottica sold over 7 million AI glasses, compared with 2 million in 2023 and 2024 combined.

Meta has led the way in smart glasses, but the basic concept is easily replicable. Numerous companies with unfamiliar names, mostly hailing from China and elsewhere in Asia, have jumped on the AI-powered smart glasses bandwagon. Audio-only glasses can cost as little as $100. Others priced at a few hundred dollars offer cameras and a display. At this stage of development, there are a significant number of companies making and selling smart glasses that claim to offer some form of AR functionality. To do this, they offload most of the computing work to a tethered smartphone.

In September 2024, Meta introduced its Orion-branded 'true' AR glasses to the world. Orion is not yet a product, but a prototype of what Meta thinks it can achieve in the next few years. It embodies cutting-edge technology that would render a production item far too expensive at this stage. The challenge is to improve the aesthetics and get the build cost down. The company has given no timescale for Orion's commercial introduction, but unofficial briefings indicate a target date of 2027. Not to be outdone by its much larger competitor, Snap intends to offer Specs to the world at large during 2026. From the little that has been revealed, these will have a large AI component and deliver some of the capabilities that Orion is seeking to offer in a commercial device.

As ever, there is minimal information on what Apple is planning. It has been active in this field for over a decade. In the absence of hard information, rumours abound. One of the more credible of these suggests that, in response to the success of Meta Ray-Bans, Apple has accelerated development of its own smart glasses. Reportedly, there are two versions in the pipeline – with and without a built-in display – due for launch in 2027.

Google is developing Android XR, which it hopes will become a universal XR operating system that it and others will use as a software platform. Android XR is heavily integrated with Google's AI. It is also collaborating with Samsung on introducing headsets and smart glasses, running on Android XR. The company has set its sights on the market opened up by Ray-Ban Meta glasses. In May 2025, it announced it was teaming up with three well-known brands in this field, including taking a stake in the South Korean fashion eyewear brand, Gentle Monster. Several smart glasses are in the pipeline from other firms using the Google Android XR platform as the base for their own offerings. The first of these, Xreal, a Chinese startup backed by the giant internet company, Alibaba, is scheduled to release lightweight glasses running on Android XR in 2026.

It is important to note in passing that American firms in the XR space are heavily reliant on China for the manufacture of their devices. Meta Ray-Bans are made in China and, from 2027, will also be made in Italy. The Vision Pro is assembled by Luxshare Precision Industry, which has become Apple's key manufacturing partner in Asia. The Chinese company GoerTek controls a major share of the VR/AR terminal market, providing headset and smart glasses manufacturing services to Meta, Sony, Samsung, Huawei, Xiaomi and others.

In sum, XR comprises a spread of technologies. At one end of the spectrum reside head-mounted displays (HMDs), as the industry likes to call them, offering the full VR/AR experience, with Apple's Vision Pro as the exemplar. At the opposite end are smart glasses offering basic photo, connective and AI functions, with Meta Ray-Ban glasses as the poster child. Ideally, a consumer device that could replace the smartphone lies somewhere in between but technological advance will not permit this to be achieved for several years hence.

25.

Keeping the show on the road 3: management, finances and strategy

MANAGEMENT CHANGES

The years after 2019 witnessed some important changes in the composition of the Cambridge Mechatronics board that, in combination, gave the company a sharper techno-commercial edge. The first of these appointments was Jorgen Lantto, recruited to be a non-executive director at the end of 2020. This was a significant move because he had recent experience designing for, and selling to, the Chinese smartphone industry. He had been chief technology officer and then CEO of Fingerprint Cards (FPC), a quoted Swedish company.

From a low base in 2009, FPC's sales rose rapidly as it carved out a position as a major supplier of fingerprint sensors to Android phone producers. Everything came together in 2016. Revenues tripled and the share price increased more than tenfold. The annual report estimated that 50–60 per cent of Android smartphones included the company's fingerprint sensor in their handsets in that year. Jorgen decided to leave FPC in 2016 and to pursue a portfolio career as a director at companies active in technologies allied to his existing expertise, selling the majority of his shareholding when he left. He is now on the board of five such companies, including CML where his experience of dealing with Chinese smartphone customers is clearly germane.

The expertise available to CML at board level was further boosted in July 2021 by the appointment of Dr Raj Talluri as a non-executive director. After 16 years at Texas Instruments, nine years at Qualcomm and four years at Micron Technology he can justly lay claim to be regarded as silicon aristocracy. Raj's experience encompasses executive roles spanning business, engineering management and strategic marketing. In January 2023, he left Micron to become president and CEO of Enovix Corporation, listed since the second quarter of 2021 on Nasdaq. Modern power-hungry portable devices are increasingly constrained by battery technology. Enovix's belief is that it can overcome these constraints through a novel technology it has developed and patented. The smartphone market is front and centre of Enovix's marketing strategy.

A third key appointment was Andrew Bickley, in January 2022, as global sales director. Although Andrew has a degree in electrical engineering, he had spent his entire 24-year career in the electronics industry in sales and marketing roles. His arrival brought new energy to the selling function. The difference was soon apparent. In January 2023 he was promoted to managing director of global sales, marketing and engineering, with a seat on the board.

Finally, in October 2022, after ten eventful years at the helm, Martin Knight stepped down as chairman of Cambridge Mechatronics. Jorgan Lantto, who had been a non-executive director for two years and was now familiar with the company, took over the chairman's role. Martin remained a director.

R&D AND FINANCES

The years after 2019 at CML can be characterised as 'the great leap forward'. 'Leap' refers to investment in R&D – not unfortunately recovery of the cost of the leap via revenues. Losses continued to be made. Despite this, as we have seen, CML perceived that the company had no lack of opportunities and was able to convince investors to give it the wherewithal to pursue them. The table below (from CML's annual reports) shows the trend.

YEAR	TURNOVER £MN	R&D EXPENDITURE £MN	NEW MONEY RAISED £MN
2019	6.6	7.4	3.9
2020	5.4	8.2	5.5
2021	5.5	8.8	2.3
2022	10.0	10.2	6.8

CML's annual R&D expenditure more than doubled over this period, from £4.7mn in 2018 to £10.2mn in 2022. The employee count grew, to more than 120. By 2022 CML's annual cost base was running at £16mn per annum. In 2023, CML decided it was spreading its R&D too thinly. Consequently, no more resources were allocated to four initiatives: 4-wire actuators, new autofocus actuators, haptics and some of the more far out possible applications of SMA.

Over the four years to 2022 CML raised additional equity amounting to close to £20mn. Where did this come from? Increasingly, it came from corporate investors – companies associated with CML as customers, licensees, suppliers and collaborators. The way it happened was that, in the middle of 2020, two of the firms in CML's orbit indicated that they would be interested in buying a small stake in the business at the next fund-raising. Quite apart from the monies involved, this was welcome news to CML, in the belief that partners with a vested interest in the company would be more likely to incline their commercial policies in its direction.

This became reality in 2021 when Huawei's venture capital arm, Hubble, and SAES Getters joined the share register via the purchase of Preferred Ordinary Shares, first issued in 2018. Encouraged by this injection of corporate cash, CML put out feelers to others in the industry to gauge further interest. All in all, CML talked to a dozen companies – all names familiar to readers of this book. These conversations bore fruit in 2022, when Alps Alpine and GalaxyCore both invested in the Preferred Ordinary Shares. Later that year they were joined by TDK.

An obvious question arises from these transactions: what was in it for the corporates? Why did they choose to invest? Each investor had a different set of priorities, but certain common threads can be surmised. Generally, the desire to make a profit on CML's exit would have been secondary to the right to an inside track on what CML might be planning. Similarly, investors in CML would, in a supply crunch, have leverage over CML and its licensees in regard to obtaining deliveries. But, above all else, a small equity stake grants the investor with a seat at the table when the future of the company is under discussion.

PATENTS AND NEW HORIZONS

The increased commitment to R&D following the re-financing can be seen in the subsequent surge in patenting activity. Since 2019 CML has been within the Top 50 companies for UK patent applications, as assessed annually by the government-owned Intellectual Property Office. Patenting activity was not, of course, confined to consumer electronics but also directed towards other potential uses of SMA. This process was given new impetus in 2021 with the appointment of a business development director (new products) with a brief to intensify these efforts.

One of these diversification initiatives came to fruition in June 2022 with the announcement of a collaboration with a long-established private Italian group Fluid-o-Tech, to apply SMA actuation to its range of fluid control products. Fluid-o-Tech concentrates on digitally controlled precision valves to mix, dose and sense fluids. Since 2014 the company had been using SMA actuators of its own design to control valves in a variety of medical and industrial settings. What CML could contribute was its reservoir of expertise in actuator control software. It was a chance to extract a high margin from the IP embodied in the firmware, representing many years of software creation.

OIS DOES THE SPLITS

By the beginning of 2024, CML had cumulatively delivered 75mn primary camera actuators. OIS remained *de rigeur* for the primary camera on premium models, with an ambition to move down the price scale. As discussed in Chapter 23, there had developed a parallel requirement for the addition of OIS to telephoto lenses. Nevertheless, primary camera modules remained the dominant market, accounting for around 75 per cent of all OIS actuators sold annually.

Since 2019, the race had been on to apply OIS to periscope cameras. On the face of it, it would seem logical for the established primary camera OIS actuator suppliers to provide actuators for periscope cameras. Unfortunately, technically this was not possible. The nature of the movement demanded by a periscope camera actuator is quite different from that required to move the lens (or the image sensor) in a primary camera. Put simply, SMA actuators work well where the necessary movement (the 'travel') is small but the force required to move the lens is large. What was needed was the reverse of this: an actuator mechanism with a long travel able to move lighter lenses in a telephoto camera. Once again, VCM showed itself capable of being stretched into providing an adequate solution.

FINANCIAL CONUNDRUM LEADS TO SILICON STRATEGY

Management had been thinking for a while about how the company's business model could be adjusted to improve financial returns. Clearly, the advantage of a licensing model is avoidance of the need to set up manufacture with all the associated costs. The countervailing downside is the lack of control over licensees and the inability to capture the manufacturing margin that accrues to the licensee. Typically, a 4-wire OIS actuator sold for $3–4, which is less than 1 per cent of the cost of making a premium phone. CML's return for the IP and the assistance given to licensees was a mere 15–20 US cents per actuator sold. This was the structure that had been established from the beginning. It was not easy to persuade licensees that the royalty rate should increase.

To achieve financial viability through royalty payments CML needed its licensees to be shifting at least 150mn actuators per annum. With the current market structure this was simply not feasible. Why so? The numbers in 2023, according to Counterpoint Research, stacked up like this:

- More than 700mn OIS actuators were manufactured in 2023.
- Most premium phones, notably the iPhone and Galaxy ranges, offered two cameras with OIS: the main camera and the telephoto. OIS was rarely added to other camera modules.

- Of the approximately 1.1bn handsets shipped in 2023, about 30 per cent contained OIS actuators.
- However, demand for OIS actuators was growing steadily. During Covid and beyond, sales of premium phones continued to do well relative to lower-cost devices.
- On the basis of 700mn OIS actuators sold to smartphone customers, the 22mn units manufactured by CML's licensees in 2023 implied a market share of 3 per cent.
- The market would not see a step function increase in demand until OIS actuators were cheap enough to go into mid-range handsets.

The inescapable fact facing CML was that the great bulk of this annual output of OIS actuators was destined for upmarket iPhones and Galaxy handsets. Hence the imperative to – somehow – gain access to one of these two companies to achieve the necessary volume. There *is* a market for OIS among the Chinese phone makers, but it is a residual one and manifestly insufficient to create a sustainable business.

CML looked for a way out of this conundrum and realised that an adjacent area offered the potential for raising the level of financial return: driver ICs. Driver ICs are supplied, for the most part, by specialist semiconductor companies and are purchased directly by the camera module integrator from the manufacturer. However, a close working relationship between the actuator designer and the actuator driver manufacturer is essential to ensure compatibility. After all, the driver IC contains firmware written by CML specifically for that actuator. In essence, CML felt that driver IC makers had captured a nice margin on the back of CML's software capabilities. Management began to explore ways in which the company could access that margin.

A handful of fabless companies specialise in what is a minor segment of the semiconductor market. Camera IC driver chips *can* be bought off-the-shelf by actuator makers and programmed accordingly.

Some larger semiconductor firms, such as Renesas and Rohm (both Japanese) and Onsemi of the USA, supply smartphone camera control chips but it is of marginal importance to them. For a long time, CML made do by buying driver ICs designed for VCM actuators from these companies. A dedicated ASIC chip designed to work with CML's own actuator is a far better solution.

Apple, characteristically, eschews the IC driver manufacturers mentioned above. Instead, it uses Cirrus Logic and Analog Devices, two US-based semiconductor companies. Most of the iPhone's camera controllers – the term it prefers to use rather than drivers – come from Cirrus. Over 15 years, Cirrus has built a highly successful business, initially on the back of the supply of audio chips for the iPhone. The company's reliance on Apple has climbed to over 80 per cent of revenues. Cirrus's stock price has more than quadrupled in the last ten years. It undergoes periodic bouts of nervousness as investors confront the unwelcome prospect of a change in Apple's purchasing policy.

Looking to create its own driver IC, in 2020, CML turned its attention to smaller (fabless) semiconductor companies for whom driver chips are a significant business line. One of these was Fitipower Integrated Technology of Taiwan, yet another tentacle of the sprawling Foxconn/Hon Hai group. Quoted on the Taiwan Stock Exchange, Fitipower's original business focused on driver ICs for displays used in laptops, monitors, PCs and other consumer electronic products, so a move into designing driver ICs for smartphone cameras was seen as a logical extension. The company was contracted to design an ASIC to work in conjunction with CML's 4-wire OIS actuators.

CML's silicon strategy was so called because the objective was to exert greater influence over, and extract more value from, the semiconductor chip that accompanied every actuator. Crucially, the 'system cost' – as the company described the package – would be cheaper for the customer.

Silicon was a new world for CML, albeit a familiar one from a distance. In 2021 CML came up with a four-phase entry plan.

1. **Royalty**. From the first half of 2021 CML began to receive what it referred to as a 'firmware royalty' from each chip sold by the driver IC suppliers to the camera module integrator.
2. **Branded**. This involved buying IC driver chips, stamping CML on them and then delivering them directly to the camera module integrator, thus gaining a margin.
3. **Co-developed**. CML partners with a semiconductor supplier to develop driver chips for sale under its own brand.
4. **Fabless**. CML funds development of a chip entirely to its own design, thereby accruing a large slice of the silicon supplier's margin.

Creating a complex ASIC is a time-consuming job with hurdles along the way. It is common for a design to go through several iterations before being sent to a foundry for fabrication. Fitipower was contracted to design an ASIC (CM824) in pursuit of phase 3. In March 2022, an agreement was signed with a small, UK-based designer of ASICs called Sondrel to create a more advanced driver IC (so reaching phase 4) specifically for the 8-wire actuator (CM825). Conscious of the geopolitical risks of relying on a single source of driver ICs and the foundry that fabricated it, CML resolved that it would utilise these products in parallel.

The May 2025 Shareholder Update contained the following helpful diagram summarising how the new business model would operate when this second project came to fruition, with a solid revenue stream flowing from the camera module integrator to CML as a result of controller (driver) IC and firmware sales.

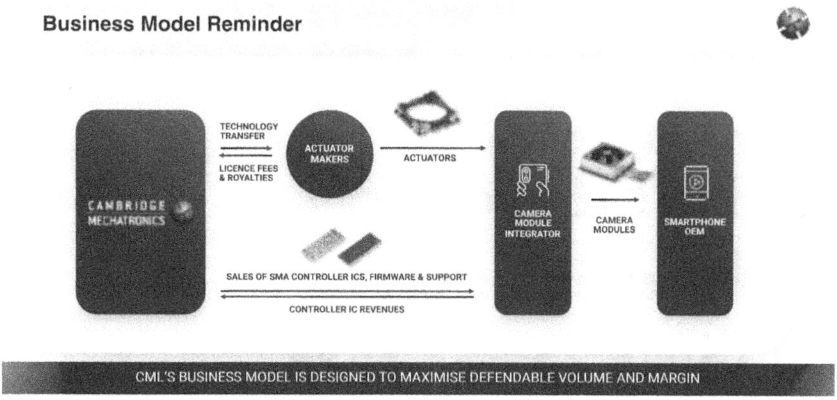

Source: Cambridge Mechatronics

CML was relying on these initiatives to provide the boost in profitability it sorely needed during the latter part of 2022 and subsequently. However, the company recognised that there was a further hurdle that could materially extend the timescale. Customers insisted on a rigorous qualification programme for these new driver chips before agreeing to their insertion alongside their companion SMA actuator.

26.

China's drive for self-sufficiency, funding to the fore

OBJECTIVE: SELF-RELIANCE

The imposition of sanctions on Huawei in 2019 was perceived by the Chinese authorities as a frontal assault on a beacon of Chinese technology. Huawei lay at the centre of China's determination to control the technology and make domestically as many of the components needed for smartphones (and other equipment) as possible. All of this had obvious potential negative implications for CML. State funding poured into electronic components, with semiconductors at the centre of this thrust.

This accelerated policy of supporting local manufacture should be viewed in the context of China's long-term industrial ambitions. These were first codified in 2015 – before any overt Sino–American friction – and incorporated into what was designated as the Made in China 2025 (MIC 2025) policy. According to a paper published by a Stockholm-based research and policy organisation:

> 'MIC 2025 is an initiative which strives to secure China's position [as] a global powerhouse in high-tech industries. The aim

is to reduce China's reliance on foreign technology imports and invest heavily in its own innovations in order to create Chinese companies that can compete both domestically and globally.' ('Made in China 2025: Backgrounder', Institute for Security & Development Policy, June 2018)

Despite the relatively benign geopolitical environment at the time, MIC 2025 did raise concerns in the USA and among Western nations generally.

When first conceived, Made in China 2025 was more of an aspiration than a plan. The 2019 American embargo on Huawei understandably injected a sense of urgency. Resources were mustered and Huawei became the epicentre of a web of government support measures, both financial and practical. Huawei's critical problem lay in semiconductors so, to assist the company, the government resolved to construct a fully home-grown semiconductor supply ecosystem. The focus was on backing chip startups, with the intention of replacing those elements in the supply chain that relied on foreign technology with domestic solutions.

What seems to have happened is that Huawei initially toyed with the idea of building its own state-of-the-art wafer facilities. Instead, it was 'encouraged' to collaborate with SMIC, the leading Chinese foundry, listed on Shanghai's STAR stock market. SMIC appeared to offer the best chance of being able to create a domestic source of Kirin SoCs for Huawei's flagship phones. However, it too was hobbled by sanctions. In December 2020, the USA declared SMIC to be a 'military end-user' and blacklisted the company, curbing expansion by restrictions on the export of critical semiconductor manufacturing equipment and Electronic Design Automation (EDA) tools. In spite of this, SMIC embarked on a major wafer fab building programme. Backed by central government and quasi-government institutions, finance was not a constraint. SMIC raised $6.6bn on entering Shanghai's STAR market in July 2020. Capital expenditure has been on a high level since then, running at $7bn annually.

Consistent with the Made in China 2025 policy, Huawei has adopted a strategy of investing in small Chinese companies developing and making semiconductor-related components or equipment. In 2019, Huawei set up two separate corporate venture capital vehicles, Habo and Hubble, with broadly the same remit: to invest primarily in companies capable of providing the technologies that US sanctions had denied it. The focus was on upstream capabilities, such as specialised manufacturing and testing equipment, EDA tools and certain input materials. The two CVCs are structured differently. Habo is an in-house operation while Hubble is organised as a venture capital partnership, with co-investors. Numbers are not easily obtainable, but it is believed that each had bought stakes in approximately 100 companies by 2024.

The graphic below is an assessment of the level of government support worldwide that has gone towards expanding local semiconductor capacity.

Figure 1. Global Chip Sector Financial Support, $bn

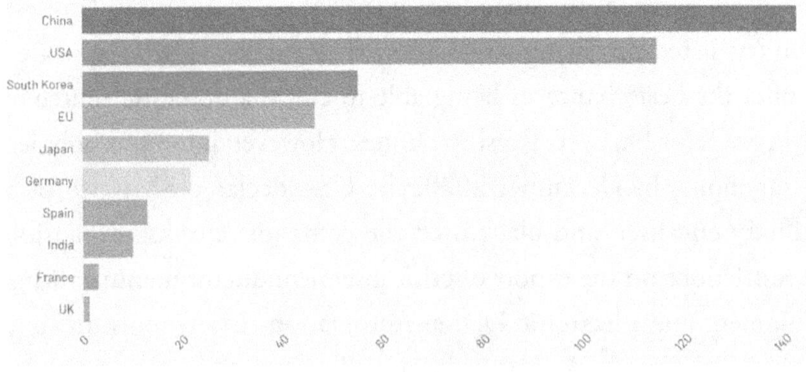

Source: Hawkins et. al., 2024 in 'Towards a data-driven UK semiconductor strategy', ESRC Centre for Digital Futures at Work, 2025

What stands out immediately is that, in terms of the sheer volume of funding, led by the National Integrated Circuit Industry Investment Fund (the Big Fund), China is in the lead. It is important to bear in mind, though, that most of this investment is for less advanced chips. Moreover, there are strong indications that these funds have been not dispensed efficiently. In 2022 the country's top anti-corruption agency launched investigations into various groups associated with the Big Fund. This was followed in March 2023 by a tacit acknowledgement that ill-targeted subsidisation of the domestic semiconductor industry had produced poor results. Chinese government policy shifted to more focused support, to be channelled to a select group of chip-related companies.

The Huawei Mate 60, with CML's actuator inside, reflected the extent to which China had reached its goal of self-sufficiency. Considerable progress had been made. According to the Japanese technical consulting firm, Fomalhaut Techno Solutions, the value of domestic-sourced components in the Mate 60 Pro comprised 47 per cent of the total (compared with 29 per cent for the Mate 40 two years before). American and Japanese components had been virtually squeezed out. The only country to figure large, other than China, in the Mate 60 supply chain was South Korea (36 per cent). From the Chinese perspective, reliance on South Korea – over which it believes it has some political leverage via its relationship with North Korea – was presumably considered to be the least-worst option.

On the surface, Huawei's smartphone resurrection appeared to be unequivocally positive for CML. However, China's relentless pursuit of supply-chain self-sufficiency left CML – as an island of foreign technology within a flagship smartphone – in an exposed position. Conscious that the long reach of the USA could further tighten the sanctions screw at any time, Huawei told CML that it intended to substitute a wholly Chinese designed and made VCM actuator for the 8-wire actuator in the forthcoming P70.

This was unwelcome news, but CML was assured that the replacement was a temporary measure – and that SMA OIS would be reintroduced in the upcoming Mate 70 and subsequent models. It was assumed that, guided by government, the intention was to demonstrate that a Chinese part *could* be substituted. On a personal level, the relationship with the executives at Huawei that the CML team had worked with for years remained good. Naturally there was concern but, as a provider of a low-profile, seemingly medium-tech, item, CML felt it had a chance of escaping the on-going cull of non-Chinese inputs.

SILICON STRATEGY AND FURTHER DEVELOPMENTS

What of CML's silicon strategy? Specifically, the supply of CML's first self-developed actuator driver chip by Fitipower (CM824). To reach the point of delivery, it was necessary for CML to meet rigorous qualification procedures to get these units onto Huawei's (and Honor's) 'Approved Vendor' list. Huawei pushed hard for this to happen, with the intention of inserting the 8-wire actuator together with the new driver chip in the Mate 60 due later in 2023. It was touch and go, due to technical issues at Fitipower, whether the driver IC would be ready in time.

From a financial perspective, silicon development, in the form of payments to Fitipower and Sondrel, exerted a considerable drag on CML's bottom line, to the tune of £2mn in 2023. Admittedly, revenue was beginning to come in from the 'firmware royalty' levied on driver IC sales but it was miniscule. Disappointingly, the CM824 did not meet the deadline for the Mate 60.

Of all the new developments that CML was working on, module tilt appeared to offer the most promise. The engineering team at CML was convinced that the technology they had invented represented a step function improvement over the competition in terms of video stabilisation. The reaction of key potential customers appeared to confirm this. Samples were delivered to Huawei in January 2023.

Camera module engineers are engaged in a constant quest to emulate what a stand-alone digital camera can do inside the constricted space of a handset. Variable aperture (frequently abbreviated to VA) is just such a feature. The lens of a digital camera consists of a series of aperture blades that form a circular opening. These blades can be widened or narrowed by an actuator to let in more or less light. Pictures taken with VA look sharper, have less image noise and handle extreme lighting situations better.

Samsung introduced variable aperture in the Galaxy S9 back in 2018. It appeared the next year in the S10 but had vanished by the time of the S20. It was then revived by Huawei and Xiaomi. This renaissance was prompted by the widespread use of larger sensors and bigger lenses. Large lens apertures can cause compromises in picture and video quality when taking images in well-lit conditions. Variable aperture ensures there is no loss in photographic quality, when shooting video especially. The few handsets incorporating VA at this stage used VCM to drive the lens-moving mechanism. CML immediately perceived an opportunity here, taking advantage of SMA's smaller size and lower weight, among other benefits.

Foldable phones first appeared in 2018. They are either flip-type (reprising the clamshell handsets of the 1990s) or book-type. Among the mainstream manufacturers, the Samsung Galaxy Fold at the end of 2019 was the first folding device that achieved recognition on a wide stage. Samsung was soon being pursued by all the Chinese manufacturers, led by Huawei. Foldables were perceived as an opportunity for CML. To maintain thinness the camera module needed to have a low height profile, something that SMA can offer. Folding phones use magnets to ensure the sides of the device are securely held together when closed, which means VCM actuators embodied in a foldable device need to be carefully designed to avoid magnetic interference. No such consideration needed to be given to an SMA-based actuator.

Book-type phones that fold have a distinct appeal for those who want on-the-move access to a larger screen, but sales have been inhibited by the high price. By 2025, however, they constituted a fast-expanding niche market, while accounting for less than 2 per cent of global smartphone shipments. Folding devices are particularly favoured by high-end consumers in China and, accordingly, the country accounts for the bulk of worldwide sales, with Huawei in the lead.

CML enjoyed some success in securing design wins from Xiaomi, Huawei and Honor. In September 2024, Huawei took the foldable concept a stage further with the introduction of a tri-fold phone, the Mate XT (incorporating an SMA OIS actuator). Huawei's ability to create such a complex product caused a stir worldwide. In China, the announcement of the Mate XT was deliberately timed to upstage the launch of the iPhone 16, flaunting Huawei's engineering prowess in the face of American restrictions.

Meanwhile, GalaxyCore's ingenious idea of creating a low-cost image sensor using SMA had been beset by technical difficulties and had, over three years, gone nowhere. Xiaomi had been signed up as the launch customer, but nothing had emerged. One problem was a bad judgement call on the resolution of the image sensor. The Chinese mainstream market had settled on a 50MP sensor rather than the 100MP sensor envisaged by Xiaomi. CML could do nothing but watch from the sidelines. The 'lost' projected revenue from 2022 onwards was a material financial negative.

NEW MONEY AND NEW BLOOD

Cash, as ever, was a persistent concern at CML. Fortunately, more corporates had expressed their confidence in the company's future by investing equity amounts of between £1mn and £3mn apiece. Together with further cash injections from existing private investors, these sums plugged the gaps. By the beginning of 2023, CML had recruited six trade investors to the share register: Huawei, Alps Alpine, GalaxyCore, SAES Getters, TDK and Snap. Between them, they controlled roughly 6 per cent of the equity.

CML's 'cash runway' was projected to extend to the end of 2023. Beyond then, it was increasingly evident that, with the heavy investment in designing and making its own driver ICs and the delays in getting the 8-wire actuator into production, the company would need to raise a far larger sum than it had in the past. The board began to consider how this might be achieved. Not being privy to the actual discussions that took place, what follows is surmise.

CML had benefited for several years from a good run of financings from a combination of current shareholders, additional wealthy individuals and corporates. It appears that the board came to the conclusion that reliance on this hand-to-mouth funding route was no longer tenable. To realise its ambitions, it was necessary for the company to be elevated onto a higher plane. Securing venture capital investors with expertise in semiconductors and allied technologies, together with the industry contacts and credibility they brought with them, would lift CML to the next level. It was, of course, more than a little ironic that CML, having resisted venture capital investment from the outset, should now – close to 30 years on – accept that this was the way forward. Equally, it is worth pointing out that, had CML been funded by VCs from the start of the smartphone project, it is most unlikely that, given the normal venture capital time horizon of 10–12 years, they would have been prepared to continue to support the firm when it encountered setbacks after 2015.

Preliminary talks were held with a venture capital firm called Atlantic Bridge, which had been introduced to CML. With its roots in Dublin and an office in Palo Alto, several of the partners of Atlantic Bridge had 30 years of experience of investing in technology firms, specialising in advanced manufacturing and semiconductors. By March 2023, in parallel with discussions with Atlantic Bridge, the decision was made by the board to appoint advisers, whose job it would be to seek out potential VC and trade investors. In doing so, they would pitch CML, channelling the IC driver chip expertise that it had acquired, as a fabless semiconductor company.

Confirmation of Atlantic Bridge's willingness to commit came quickly, in the form of a term sheet – a non-binding agreement outlining the basic terms and conditions of an investment. The CML proposition was clearly one they were prepared to back, which was manifestly good news. The downside, however, was the share price that Atlantic Bridge was prepared to pay. It was well below the price paid in recent funding rounds by both existing shareholders and trade investors.

For the CML board this came as a disappointment but not a surprise. Venture capital fund-raising by US firms hit a six-year low in 2023, due to economic uncertainty and low IPO activity. Equally, VC firms were sitting on reserves of cash – giving them the firepower to drive a hard bargain. Conversely, CML, with its increasingly urgent need to refinance, was in a weak negotiating position.

In terms of the response from potential VC investors to the soundings by the financial advisers, it was heartening that other firms of stature and 'domain knowledge' (the phrase that VC firms like to use) in this sector were prepared to follow Atlantic Bridge's lead. Most notable among these was Intel Capital, the venture capital arm of Intel Corporation. Intel Capital is considered to be the *crème de la crème* within the community of high-tech VC investors, with a large portfolio of companies under its wing and extensive technical and commercial resources. There was a definite appetite for CML's equity – at the 'right' valuation!

Additional investors were recruited and, by July 2023, a consortium of VC firms, co-led by Atlantic Bridge and Intel, had been put together. Due diligence was conducted over the summer. By the time all the legalities had been completed, it was uncomfortably close to the end of the year before the deal was done. CML and its shareholders could breathe a sigh of relief. One of the advisers marked the occasion with the following graphic.

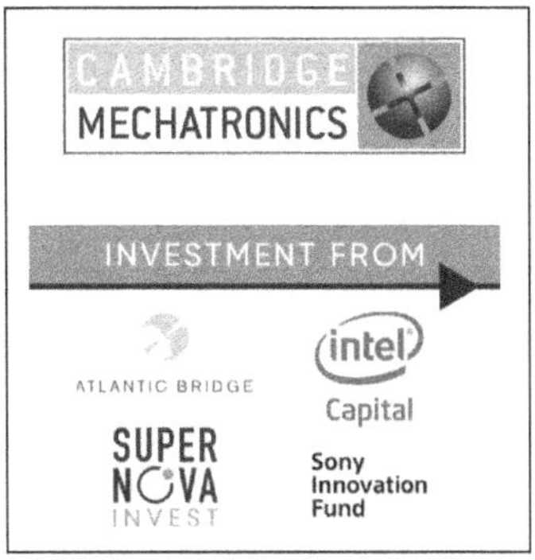

Source: Acuity Advisors

In all, by early 2024, this VC-led transaction had raised approximately £35mn, far more than had been raised in any single funding round previously. However, to secure this sum, CML had had to agree to what venture capitalists call a 'down round'. At £1.86 a share – the price of the new money – it represented a massive discount to the £6.26 and £10 a share that investors had paid not far back in time. The ensuing valuation was £92mn. On the upside, the funding round was a powerful validation of what CML had built and was trying to achieve from a bunch of hard-nosed, tech-savvy VCs.

In terms of the shareholder register post-funding, Intel Capital emerged as the largest VC investor, with just under 10 per cent of CML's equity. Second in line was the French technology-focused investor, Supernova Invest, with over 8 per cent. Atlantic Bridge ended up with close to 6 per cent while the fourth VC investor, Sony

Capital, had acquired just over 2 per cent. Broadly, four VC firms now controlled 26 per cent of CML's equity. Stewart Newton's interest in CML had been diluted to 24 per cent. A few other private investors retained substantial positions. Subsequent to the main transaction, CML received £500,000 from a seventh corporate investor, the US firm, Cirrus Logic, with which it had been in contact for several years. The relationship had developed due to several overlaps in the two companies' product lines and ambitions.

As indicated above, this transaction was more than just a matter of raising money. Partnering with top-notch VC houses, especially one that is an offshoot of a major tech firm, contributes much more than finance, in the form of tangible technical and commercial assistance to portfolio companies. Founded in 1991, Intel Capital's primary mission is to invest in and support innovative technology startups and growth companies that 'align with Intel's strategic interests'. In practice, the 'align' criterion is interpreted generously. In addition to finance – and possibly of greater import – Intel Capital offers mentorship, expertise and access to Intel's vast network of industry connections.

Supernova too has impeccable technology credentials and is backed by a slew of French blue-chip investment institutions. Sony Innovation Fund (also known as Sony Ventures) is the venture capital arm of Sony Corporation. It has been in business since 2016, with a brief to invest in sectors in which Sony is active. The established commercial relationship between Sony Corporation's image sensor division and CML was clearly a factor in the decision. Like Intel Capital, the Innovation Fund offers technical expertise and networking opportunities to portfolio companies.

The three leading investors in the consortium (Intel, Atlantic Bridge and Supernova) insisted on board representation. A hands-on approach to investments is commonplace in the venture capital world.

Three directors were appointed in December 2023, replacing some of the existing directors. The newly appointed board members were:

» Abdul Guefor. Managing director of Intel Capital's Europe, Middle East and Africa region, leading a team of 20. Based in London.
» Regis Saleur. One of the two managing partners of Supernova Invest, responsible for a 20-strong team. Based in Paris.
» Osman Kent. One of three venture partners at Atlantic Bridge. Based in London.

Securing the backing of VC firms steeped in semiconductor and related technologies was a major step forward, offering CML a reinvigorated platform for growth. For shareholders and employees, it was gratifying to read how the news was presented on the Intel Capital website:

> 'We believe CML is at an inflection point where it can both deploy its proven technology in more phone models, as well as addressing adjacent markets such as precision drug-delivery devices, XR and haptics.
>
> 'We're therefore thrilled to announce that Intel Capital co-led CML's new oversubscribed financing round, which raised over $40M. We believe we and our fellow investors (all of whom are well versed in semiconductor deals) can help CML scale its world-class technology more rapidly and build a deep tech world leader.' ('Disrupting Camera Performance with Shape Memory Alloy: Our Investment in CML', Arun Chetty and Abdul Guefor, 13 February 2024)

Intel Capital's upbeat assessment was a pleasing coda to an eventful 2023. CML had been re-energised but there was still a mountain to climb.

27.

Hanging in at Huawei, headset momentum and medical moves

ACTUATORS, DRIVER CHIPS AND MORE

In the last chapter we observed that, under pressure to localise, Huawei removed CML's SMA actuator from the rebranded Pura 70. Instead, it employed an inferior, home-grown VCM solution. With geopolitical storm clouds still gathering, there was a collective sigh of relief at CML when it was confirmed that SMA would be restored in the Mate 70, which duly appeared in November 2024. Encouragingly, the product roadmap led to the Pura 80 and Mate 80, plus design wins elsewhere in the product range. Set against this, though, was the fact that, in the first half of 2025, Huawei's offtake showed no growth.

The relationship with Huawei remained good. Nevertheless, longer term, there was no doubt that the geopolitical tide was turning against CML, with no let-up in the race to replace foreign inputs. Understandably, CML remained nervous. To illustrate the point, during 2024, it became clear that Huawei intended to supplant the Sony image sensors in its camera modules with a home-grown alternative, despite Sony's strong reputation in the field. SmartSens

Technology, a Chinese image sensor design house, was revealed as the probable vehicle for this move. Unlike its two larger rivals, Sony and Samsung, SmartSens is a fabless semiconductor company. Huawei has remained coy about the foundry that SmartSens will use to fabricate the sensor. Speculation has centred on SMIC and Hua Hong, China's second largest contract semiconductor maker.

Beyond the Mate 70, SMA actuators (both 8- and 4-wire) began to capture slots in other Huawei devices, notably the Nova mid-range of phones and foldable devices. Equally, CML started to make inroads at Honor, securing design wins in mainstream and foldable devices. However, in 2024, this did not translate into higher offtake. At 21mn, actuator volume remained flat.

This volume was, however, offset by the relentless competitive pressure on SMA to match VCM's manufacturing cost, reflected in the OIS prices that VCM suppliers were quoting. Total flagship system cost (actuator plus driver) had been forced down to between $2 and $3 and heading to less than $2. At CML's licensees, cost-cutting measures intensified, assisted by a substantial drop in the price of Nitinol wire. Reduced cost was a major factor in securing the design win for the Nova range.

From a financial perspective, the priority for some time had been the need to insert CML's proprietary driver IC into the actuator slots it had achieved, so as to benefit from the higher margin obtainable when the actuator is sold together with its driver chip. The Fitipower CM824 chip missed the boat for the Mate 60. To increase the flexibility and widen the appeal of the driver chip, CML recruited a group of specialised software engineers based in Copenhagen who were known to the chair of the board. Their task was to enhance the control algorithms to provide customers with more programming options. For technical reasons, Huawei used the CM824 in *some* of its product range. Reassuringly, by mid-2025, the CM824 started to have a distinctly positive impact on revenues and profitability, with estimates of 4mn units and £2.5mn in sales pencilled in for the year.

In regard to the next generation driver IC – aimed at enhanced performance and further system cost reduction – there was disappointment. We referred earlier to the contract that CML had signed with Sondrel, for the CM825, an upgraded driver IC specifically to power the 8-wire actuator. By the end of 2023 Sondrel had failed to come up with a design that worked. Nor was the corporate background encouraging. Sondrel experienced a cash flow crisis and had to be refinanced, followed by a withdrawal from the stock market. To retrieve the situation, CML recruited a 20-strong design team to work on the digital aspects of the CM825. All in all, the Sondrel debacle injected a precious 18-month delay. Mass production was scheduled for the second half of 2026.

As we noted in the last chapter, CML had high hopes of variable aperture (VA). Initially, firms were using VCM for this purpose, but SMA had advantages that CML believed would persuade smartphone designers to shift to its actuator. VA attracted some flagship interest, but take-up was slow and uneven. Then, early in 2025, rumours began to circulate that Samsung would include VA in its next top-of-the-range phone, after a five-year hiatus. At the end of the year, posts by the techno-commentariat suggested Apple would adopt VA in the forthcoming iPhone 18.

In the realm of fluidics, CML had moved on from Fluid-o-Tech and began to talk directly to companies in the drug-delivery field. These firms showed interest in what CML's actuator could do to upgrade current versions of portable infusion pumps for diabetes, attracted by CML's belief that SMA could offer a device half the size with ten times the flow precision.

HAPPENINGS IN HEADSETS

The engineering contracts for smart glasses/headsets were in pause mode during 2024. Work for the second American firm that had expressed interest in SMA for VR/AR resumed in 2025. Much speculation swirled round Apple on the direction it would take in this burgeoning area. As ever, leaks from suppliers fuelled the rumour mill but no hard evidence emerged. However, substantive clues can be found in Apple's published patent applications. Like every other major tech enterprise, Apple patents many inventions that are never employed in a commercial product. Nevertheless, these applications do indicate that actuators could well have a role to play in its future headsets. CML, naturally, took note.

An Apple patent application, filed in June 2024 and published in March 2025, refers specifically to the use of actuators to move liquid lenses in a head-mounted display, such as a future version of the Vision Pro. To accommodate users with different types of vision (nearsighted or farsighted) it envisaged tuneable liquid lenses, describing the mechanism as follows: 'Actuators and/or pump and reservoir systems may deform the lens surfaces in response to control signals …' Such a mechanism would avoid the need for prescription glasses. Tuneable lenses require high force – a property that SMA possesses.

Liquid lenses are not a new concept. They are widely used for industrial purposes, for machine vision, barcode reading and in life sciences. In principle, they have several advantages. The idea of using a liquid lens in a smartphone has been around for some time but only twice has it been tried in a commercial product. In both cases, the technology was provided by a Swiss company, NextLens, and consisted of a transparent fluid wrapped in a deformable membrane that can be shaped by pressure exerted by a Voice Coil Motor. (Appendix 3 explores the history of liquid lenses in smartphones and the current status of this endeavour.) An Apple patent application filed in 2023 made it clear that such a system would require multiple actuators.

In May 2025, the US Patent and Trademark Office published a patent application filed by Apple in November 2023 relating to a possible future version of Vision Pro that 'will provide left and right eye optical modules with the ability to automatically move with respect to each other with actuators ... The positions of the optical modules may be adjusted to accommodate different user interpupillary distances and/or to provide other suitable adjustments.' Interpupillary distance is the distance between the centres of the pupils of your eyes. Getting this right is essential for a comfortable and clear viewing experience in VR and AR devices.

Against this background, CML had high hopes for its technology. The objective was to position SMA as the go-to actuator technology for HMDs and smart glasses while the sector was still in its infancy. SMA's key competitive advantage was its power-saving capability – battery life remained a significant constraint on the wearer's usage time. To this end, CML had developed a unique technology called SMA Zero Hold Power (ZHP). VCMs require continuous power to maintain their position. An actuator with ZHP can be used as a normal SMA actuator but, once in position, ZHP mode can be activated and power switched off entirely, thus saving energy until another movement is needed. It is implemented through both mechanical design and firmware. ZHP SMA also offers small size, low weight and high force. Nor is it confined to usage in HMDs or other wearables but can also, for example, be applied to variable aperture in smartphone cameras.

CHINA'S UNRELENTING PURSUIT OF SUPPLY-CHAIN AUTONOMY

Under the Biden administration there was no dilution of the mission to prevent Huawei from gaining access to advanced chipmaking technology. None of this was enough to prevent Huawei, together with

SMIC, from fabricating the SoCs needed for its range of smartphones, albeit with a yield penalty. Simultaneously, there was no weakening of Chinese efforts to overcome the chokehold on EUV (extreme ultraviolet) by the Dutch company ASML. Early in 2025 reports emerged of research at a Huawei-linked company called SiCarrier directed at the development of an alternative method of lithography that would permit it to operate at low nanometre nodes. Later in the year, evidence emerged of a Shenzhen-based parallel project to reverse engineer ASML's EUV machine, staffed by former ASML employees.

Huawei, in fact, had since 2023 been surreptitiously engaged in a large-scale effort to replicate as much of the semiconductor supply chain as was possible on Chinese soil. Exactly what happened is shrouded in secrecy, but it appears that Huawei came to the conclusion that it could not rely on SMIC for the production of future HiSilicon-designed ICs. SMIC is, after all, primarily in the business of fabricating semiconductors for automotive and industrial applications and could not be expected to devote sufficient attention to Huawei's requirements. Either prompted by, or with the blessing of, the relevant ministry, Huawei embarked on a large-scale project to construct a network of wafer fabs dedicated to serving the company's needs. In effect, Huawei is bent upon turning itself into an integrated device manufacturer, like Samsung.

Strenuous efforts were made to keep this project out of the public eye. None of the wafer fabs constructed were formally associated with Huawei in any way. Several are at least partly state-owned. A leak to the Korean trade magazine *The Elec* in May 2025 revealed that Huawei's chip network consisted of 11 fabs, situated across several regions of China. They are capable of fabricating logic and memory chips and also act as foundries. Reportedly, at least five of these wafer fabs have the ability to manufacture ICs at 5nm nodes or lower. More to the point, the article disclosed that they were all operated by Huawei, with the company providing personnel, technology and systems.

These facilities were in addition to, and complemented, the more than 60 chip-related investments (according to a report in *Nikkei Asia* in May 2025) Huawei had made through its venture capital arms. Six semiconductor-related companies from this portfolio have gone public on the Shanghai Stock Exchange since 2022. It is noteworthy that all of them concealed the identity of their largest customer in their IPO filings. Huawei sits at the centre of a clandestine web of state-supported enterprises with a common goal: the creation of a Chinese chip supply chain. Confirmation of the scale and ambition of this project came in early 2025, according to the *Financial Times*.

> 'In February, Ren Zhengfei, founder of tech group Huawei, was among a select group of business leaders to meet Xi in Beijing. Seated opposite the Chinese leader, Ren told Xi that Huawei is spearheading a coalition of more than 2,000 companies aiming to boost Chinese semiconductor supply chain self-sufficiency to 70 per cent by 2028.' ('The vulnerabilities holding back Chinese industry', *Financial Times*, 30 June 2025)

Nor was this desire to construct an independent semiconductor supply chain confined to Huawei. As we saw in Chapter 7, Xiaomi has been intent on developing its own smartphone SoC since 2015. After spending a total of $7bn according to founder, Lei Jun, Xiaomi finally announced its own in-house chip, the XRing 01, in May 2025, intended to act as a signal to Qualcomm (with which Xiaomi has a multi-year supply agreement) and MediaTek. The company has made it clear this is a long-term project. Xiaomi's Achilles' heel is its reliance on TSMC for fabrication.

For Huawei, and for China as a whole, this concerted effort to create an effective national semiconductor ecosystem was being held back by a structural deficiency stemming from the way it had developed.

As in many other fields, national exhortation had created an excess of participants, often funded by local governments via Local Government Financing Vehicles (LGFVs). Accordingly, few of these firms had, or were capable of, reaching the critical mass to match their Western equivalents. National government's attempts to consolidate this fragmented landscape have been frustrated by resistance from local and other vested interests.

ECONOMICS AND POLITICS INTERVENE

This drive to create an expanded semiconductor capability has, however, had to operate in a less favourable political and economic environment. At the start of 2021 President Xi called for a 'common prosperity' campaign, designed to reverse growing income inequality. The immediate impact of this change in the political wind was a crackdown on the major internet companies, whose size and power represented a threat to the hegemony of the Chinese Communist Party. Financial markets were told it was their job to facilitate 'new productive forces' (semiconductors being one of these). These restrictive measures were felt broadly across the corporate sector and the stock market declined accordingly.

Simultaneously, the high-tech schism between the USA and China continued to widen. A set of injunctions in 2023 extended the remit of the China Securities Regulatory Commission to foreign stock market listings, placing more obstacles in the path of a Chinese firm seeking an overseas listing. In August 2023, the negative sentiment on the part of American institutions towards Chinese VC investment was reinforced by proposals announced by the US federal government to address an asymmetry: US firms were banned from selling sensitive technologies to firms in China while US investors were free to invest in those self-same firms. These new rules came into force at the beginning of 2025 but influenced attitudes well before then.

Also, in 2025, the government of Taiwan – having historically maintained an even-handed approach – unequivocally aligned itself with the American embargo on the fabrication of advanced ICs for Chinese companies. Taiwan blacklisted Huawei and SMIC by adding them to its own high-tech strategic entity list. Taiwanese companies must obtain licences before shipping products to any firm on this list. This move had no practical effect but affirms that control over TSMC's actions lies now at source rather than through the USA's Foreign Direct Product Rule leverage.

Political restraint was followed, in early 2022, by a marked deterioration in the Chinese economy, triggered by a bursting of the property market bubble. The Chinese attachment to real estate investment, rooted in the scant protection for tenants and the lack of a comprehensive social security system, meant that the real estate sector has been a major element in driving the nation's GNP. A sharp decline in asset values acted to slow economic activity and created a debt overhang, within LGFVs especially, that has yet to be worked through. During the pandemic, Beijing's economic priority was to get workers back into factories rather than, as in the West, to encourage the consumer to spend. President Xi's belief in frugality means that he is viscerally opposed to boosting consumption in any significant way. In the face of this reluctance to put money in consumers' pockets, China's economy remains investment- and export-led, while household consumption remains low by international standards.

The Trump tariff blizzard in the first half of 2025, principally aimed at China, threatened the entire basis of iPhone production. Apple's build-up of iPhone capacity in India was designed, when complete, to supply all of Apple's US iPhone requirements – until further tariff moves placed this approach in jeopardy. Faced with this onslaught, companies generally began to question the China Plus One strategy they had been pursuing. It is important to appreciate that a transfer to India did not free Apple from its dependence on China. Indian production of smartphones – estimated at 16 per cent of iPhone output in 2025 – still relied heavily on the import of Chinese low- and mid-tech components. India did not – as yet – possess the skills to make these items, let alone in volume, at an acceptable cost and level of quality. Lacking the deep bench of skills residing in China, it would take some considerable time for the country to get there.

In characteristic Trump fashion, in July 2025, in exchange for concessions on rare earths, the administration abruptly reversed an instruction issued only a few weeks earlier to American EDA companies to stop selling their software to Chinese groups, somewhat easing Huawei's and SMIC's path towards closing the gap. The Kirin 9020 SoC in the Pura 80 was fabricated by SMIC on a 7nm process, considered by industry observers to be three generations behind the 2nm process TSMC planned to introduce in 2026. Huawei may lag in process terms but, in the 9020, it accomplished the feat of integrating a 5G modem into its SoC, something that Apple had not yet been able to do.

A MAJOR MEDICAL OPPORTUNITY SHAPES UP

We noted earlier in this chapter that CML had concluded that drug delivery could benefit from what SMA had to offer: the force the actuator could bring to bear, its small size and low weight, plus the high degree of control embodied in the driver chip. The last decade has witnessed a revolution in the way diabetes is treated. At its heart are compact, wearable devices that deliver insulin continuously or in programmed doses through a tube without the need for repeated manual injections. Some people with type 1 diabetes use an insulin pump and a continuous glucose monitor (CGM) that talk to each other in a closed loop system. The control programme is on the user's phone or inside the pump. Doses of insulin are adjusted automatically in response to blood sugar levels, which are monitored all the time by the CGM.

What the diabetes industry refers to as 'patch pumps' are disposable devices that attach to the body with adhesives and automatically administer insulin from a built-in reservoir. Patch pumps work in conjunction with an attached blood-glucose sensor. Significantly, the system is tubeless, in contrast to the old-style method of tethered insulin infusion. The pump is controlled wirelessly by a smartphone app. This system has many advantages for the user.

The first medical device manufacturer to develop a tubeless insulin delivery system and an accompanying mobile app was Insulet. Commercialisation of Insulet's automated insulin delivery system, which it called Omnipod, occurred in 2005. Insulet's latest version, the Omnipod 5, is currently the only patch pump for type 1 and type 2 diabetes cleared by the US Food and Drug Administration (FDA). Accordingly, it currently has an effective monopoly in patch pumps. Insulet has made all the running in tubeless insulin pumps since 2015, leaving the established firms struggling in its wake.

Insulin delivery is a crowded field. Medical device companies of all shapes and sizes compete vigorously. Historically, the leader in this business, with a 70 per cent share globally, was the broad-based American medical device company, Medtronic, founded in 1949. Diabetes is only a small part of what the company does – it is also active in several other therapeutic fields. Additional contenders in the patch pump space – other than Insulet – include Tandem Diabetes Care of the USA and the Swiss firm, Ypsomed. In recent years all of the incumbents have lost market share to Insulet's pioneering device and are striving to catch up. As if to emphasise the importance of this goal, in May 2025 Medtronic announced the spinoff of the diabetes segment of its business into an independent, listed company. It will be called MiniMed, reflecting the original name of the company before

Medtronic acquired it in 2001. A major motivation for the split was the desire to create a dedicated entity better equipped to compete with an insurgent Insulet.

Currently, infusion pump technology is directed at those with type 1 diabetes, which is a life-long autoimmune condition. However, more than 90 per cent of diabetes patients have type 2 diabetes, which is surging globally due to ageing populations, and its link to sedentary lifestyles and poor diets. Medical device producers are working hard to obtain regulatory approval to extend patch pump technology to this much larger patient population. Insulin pumps do not require hardware changes to expand to type 2 but the algorithms that ensure safe and effective delivery need adjustment. Serious complications can ensue if insulin is under or over delivered. In the USA, the FDA cleared the first automated insulin delivery system for use with type 2 diabetes in 2024.

Wearable insulin pumps consist of several components housed inside a plastic frame. As well as an insulin reservoir, a plastic needle and batteries, the crucial operating mechanism comprises an actuator to drive the pump plus a control chip to deliver the correct doses. Patch pumps need to be disposed of and replaced every three days. They currently represent a market of around 60mn units a year. With extension to the large population of type 2 patients, substantial growth is in prospect. By early 2025 CML was in deep conversation with several of the larger medical device manufacturers. The company believes the pump it has designed will be half the size and weight of an Omnipod 5. Given its prospective growth, the insulin pump market represents a significant opportunity for CML. Patch pump consumables are expensive items, costing tens of dollars for three days' worth of insulin. Consequently, the company also hopes that the pressure on margins experienced on smartphone actuators will not be encountered in this market.

28.

Keeping the show on the road 4: management, finance and strategy

APPOINTMENT OF CEO

It inevitably took a few months after the refinancing in early 2024 for the reconstituted board of CML to bed down. However, it was noticeable from the outset that the presence of three directors with long experience of guiding small and growing companies served to elevate the level of discussion. On top of this, the new directors were the source of much practical help, particularly with regard to contacts at potential customers or partners. Armed with a new sense of confidence, CML moved to offices with better facilities in Cambridge Science Park. However, it soon became clear that these and other initiatives, together with the familiar lack of actuator volume plus the delay in introducing Fitipower driver chip, meant CML would incur a large loss. Measures were accordingly taken to reduce headcount.

During 2024 the VC firms now guiding the strategy at CML concluded that the company needed to move to a conventional corporate structure with a CEO at the helm, in place of the managing director/executive chair structure that had been in place for more than ten years. They embarked on a search to find an experienced CEO with a semiconductor background (reflecting the repositioning of CML as a fabless semiconductor firm). The outcome was the appointment in April 2025 of Dylan Kelly as – echoing American practice – President and CEO. As well as 15 years' experience leading small R&D-intensive US firms in semiconductor-related fields, Dylan was familiar with the smartphone industry, having started his working life there as a design engineer.

One of Dylan's first actions was to question the sharp uplift in R&D expenditure in the previous two years, fuelled by the major injection of cash in early 2024. From just over £10mn per annum in 2022, it had risen by 30 per cent in 2023 and a further 20 per cent in 2024, to reach £15.8mn. Maintaining and improving CML's proprietary technology to fend off any competition was obviously crucial, but this level of commitment risked tilting the company towards becoming a research house rather than an organisation pursuing commercial gain. This had been recognised before his arrival and actions were already in train to reduce the engineering workforce. By the second half of 2025 CML was down to 100 employees, from 160 eighteen months earlier. China remained a priority, with 13 dedicated engineers located there.

INNOVATION AND STRATEGY

CML's assiduous research efforts to expand the boundaries of what SMA could do had spawned an array of actuator solutions for the different camera sockets on a smartphone. To an outside observer, CML's product range now consisted of a bewildering array of 'horses for courses' actuator variants. In addition to telephoto/periscope cameras, ultra-wide and front-facing cameras were also emerging as candidates for OIS. However, positivity surrounding these further opportunities should not be allowed to obscure harsh, commercial reality. In the foreseeable future, CML's fortunes were inextricably tied to its ability to penetrate the dominant market for actuators on phones – the primary camera – with its 'workhorse' 8-wire AF+OIS product.

A review of the progress made since 2010 is instructive. The diagram below shows how far SMA actuator and control technology had come since the original autofocus device.

SMA actuator development milestones

Introduction date	
2010	SMA lens autofocus (AF) actuator (2-wire)
2015	SMA lens OIS actuator (4-wire)
2021	SMA lens AF+OIS actuator (8-wire)
2022	SMA sensor shift AF+OIS actuator (8-wire)
2023	SMA module tilt OIS
2023	SMA variable aperture
2024	SMA Zero Hold Power (ZHP)

Source: Tony Golding

Overall, in mid-2025, CML had firmed up on a three-pronged strategy:

- The company intensified its endeavours to persuade smartphone manufacturers to adopt SMA OIS (conscious that sales to its major customer rested on an insecure foundation).
- The contract engineering work that CML had been engaged in since 2022 was evolving into a considerable opportunity for SMA actuators in VR/AR head-mounted displays and smart glasses.
- Portable, disposable patch pumps for diabetes patients were shaping up to become a significant new market for CML's actuator and control system expertise.

Postscript

A FEW OBSERVATIONS ON CML'S EXPERIENCE

» In a complex, multi-level supply-chain environment, setbacks in a small company's quest to insert its product into the marketplace can come in unexpected ways and from unexpected directions. In the smartphone industry, for a company to secure a design slot for its component on a new phone, every piece in the supply-chain jigsaw has to fall into place. A delay or failure anywhere in the network of suppliers will have fatal consequences for the innovator's project.

» In this environment it is crucial for an innovating startup to sound out the right partner or partners – whether in the form of customer, supplier or collaborator. Conversations with industry participants take place constantly in an effort to find common ground on which a mutually advantageous relationship can be built. Most discussions lead nowhere. Then, as the story of CML demonstrates, they can resume, maybe after a long break. Everything takes longer than anticipated and most initiatives run into sand.

» CML's progress has been halted or interrupted numerous times by an astonishingly diverse set of circumstances, often wholly or largely outside the company's control. Above all else, after 2019, Sino–American tensions dislodged CML from the growth trajectory on which it was set.

- That CML has coped as well as it has with these vicissitudes is a tribute to the members of the core engineering team that has held together since the beginning. This is an often-overlooked attribute of startups that endure: the founding team retains its coherence, respecting each other and working well together.
- Until 2023 CML was financed by regular infusions of cash from private individuals. It is arguable that it would have been better advised to have sought the larger sums available from venture capital sources at an earlier stage. Whether that would have been possible in the 2010s, given the absence of an SMA breakthrough and lacklustre sales, is, however, an open question.
- Around 20 companies, both new and established, have also striven to replace the incumbent smartphone VCM camera component with something better, employing a variety of technological approaches. The plain fact is that none has, so far, succeeded. Appendix 3 (see www.downtothewirecont.com) surveys these competing approaches and their commercial outcomes. Their experience is instructive and serves to confirm that what CML encountered is similar to what all newcomers face.
- Finally, the ineluctable truth is that CML has been a victim of geopolitics. CML became dependent on Huawei, which, to its credit, was the only smartphone producer prepared to take a chance on an unproven technology on terms acceptable to CML. Since the second half of 2020, Huawei, has been a political and economic pariah, forcing its previously successful smartphone business to retrench. Had Huawei been able to continue to operate at the 200mn plus handset output level it achieved in 2019, there is every possibility CML would have attained financial viability by 2022.

THE CML SAGA CONTINUES …

The CML saga has not yet ended. For those who wish to discover the course of events subsequent to the completion of this book, I have set up a dedicated website at www. downtothewirecont.com. Further chapters will be posted there.

APPENDIX 1

ABBREVIATIONS USED IN THE TEXT

AF	autofocus
AI	artificial intelligence
AOSP	Android Open-Source Project
AP	Application Processor
AR	augmented reality
Arm	(originally) Acorn RISC Machine
ASG	Actuator Solutions GmbH
ASIC	application-specific integrated circuit
ASML	(originally) Advanced Semiconductor Materials Lithography
Billu	Shanghai B.L Electronics Co.
BIS	Bureau of Industry and Security
CCD	charge-coupled device
CCP	Chinese Communist Party
CEO	chief executive officer
CFO	chief financial officer
CML	Cambridge Mechatronics Limited
CMOS	complementary metal-oxide-semiconductor
CP	computational photography
CPU	Central Processing Unit
CVC	corporate venture capital
DRAM	Dynamic Random-Access Memory
DSP	Digital Signal Processor
DSP	Digital Sound Projector (brand name)
DUV	deep ultraviolet
EDA	Electronic Design Automation

EIS	Electronic Image Stabilisation
EMS	Electronic Manufacturing Services
EUV	extreme ultraviolet
FDPR	Foreign Direct Product Rule
GGIF	(Chinese) government-guided investment fund
GMS	Google Mobile Services
GNP	gross national product
GPU	Graphics Processing Unit
HDD	Hard Disk Drive
HMD	head-mounted display
HTI	Hutchinson Technology Inc.
HUD	head-up display
IC	integrated circuit
IDM	integrated device manufacturer
IoT	Internet of Things
IP	intellectual property
IPO	Initial Public Offering
ISP	Image Signal Processor
LCD	Liquid Crystal Display
LGFV	(Chinese) Local Government Financing Vehicle
LiDAR	light detection and ranging
MEMS	micro electro-mechanical system
MP	megapixel
MR	mixed reality
Nitinol	Nickel Titanium Naval Ordnance Laboratory
Nm	nanometre
NPU	Neural Processing Unit
ODM	Original Design Manufacture
OIS	Optical Image Stabilisation
OS	operating system
OSAT	Outsourced Semiconductor Assembly and Test

RAM	Random-Access Memory
ROM	Read-Only Memory
SEMCO	Samsung Electro-Mechanics
SEP	Standard Essential Patents
SMA	shape memory alloy
SMIC	Semiconductor Manufacturing International Corporation
SoC	System-on-a-Chip
SSD	solid-state drive
TDK	(originally) Tokyo Denki Kagaku Kōgyō K.K.
ToF	Time-of-Flight (camera)
TSMC	Taiwan Semiconductor Manufacturing Company
UI	user interface
VA	variable aperture
VC	venture capital
VCM	Voice Coil Motor
VR	virtual reality
XR	extended reality
ZET	Liaoning Zhonglan Technology Co.

APPENDIX 2

SMARTPHONE TIMELINE: MAJOR MILESTONES

2006	First capable smartphone (Nokia N95)
2007	Apple iPhone launched
2008	First Android smartphone (HTC Dream)
2009	4G connectivity becomes available
2009	First Samsung smartphone (Galaxy GT-1750)
2016	Google Pixel brand phones launched
2017	Multi-camera phones become standard
2018	First folding phones introduced
2019	5G connectivity becomes available
2020	Wearable devices become commonplace
2024	AI-enabled phones enter the mainstream

APPENDIX 3

THE RACE TO REPLACE VCM: A SURVEY

Scan the QR code below or visit
downtothewirecont.com/appendix-3/
to read the Appendix 3.

APPENDIX 4

SUPPLY CHAIN INTERACTIONS: CO-OPERATION AND COMPETITION

As we have seen, the smartphone supply chain is extremely complex with a labyrinthine set of relationships. Not only complex but paradoxical, in that the companies involved simultaneously co-operate (buyer–seller) and compete (at the smartphone level). This state of affairs is not unusual in the commercial world but here it is writ large. It works because it makes business sense. In short, pragmatism rules. However, tensions abound. Thirty years ago, an academic coined the word 'coopetition' to cover such relationships. Fortunately, given how singularly inelegant this literary construction is, it has not entered common usage.

This section takes a brief look at a few of the more outstanding examples of collaboration allied to competition. Relationships are often characterised by turbulent inter-dependence, akin to parents trapped in a troublesome marriage for the sake of the kids.

Apple–Samsung

Apple and Samsung have sued each other numerous times, claiming technology theft, but the companies agreed a truce in 2018. Apple experiences the most angst in this relationship as it is largely reliant on Samsung for displays. Handset competition is intense. Samsung's Galaxy S range is pitched deliberately at the iPhone. It emerged in May 2023 that Apple has plans to ensure it is not beholden to Samsung for the next generation of displays. This is not an immediate prospect, however.

Apple–Qualcomm

Apple deeply resents its reliance on Qualcomm for the wireless modem chips that all smartphone producers are obliged to buy. Qualcomm designs and sells these critical ICs and has a grip on the underlying patents. Displacing Qualcomm has not been easy. An Apple Silicon modem has been in the pipeline since 2019 following the purchase of Intel's modem business and finally appeared in a new phone in 2025. In the meantime, Apple continues to pay Qualcomm for the use of its patents *and* to purchase Qualcomm modems for the iPhone.

Samsung–Sony

Samsung is prepared to purchase externally for its smartphones if it cannot find a suitable in-house solution. Image sensors for its camera modules illustrate this policy in action. Even though Samsung has powerful capabilities in this field, flagship phones may well contain image sensors from Sony as well as from its own facilities. The Samsung Galaxy S23 Ultra, for example, used two Samsung image sensors and three from Sony. The first application of a new Samsung sensor may well be in a Xiaomi or Vivo phone rather than in a Samsung handset. Similarly, Sony uses Samsung image sensors in its own handsets. It is a question of what makes technical and commercial sense.

LG–Apple

LG's decision in 2021 to withdraw from making smartphones has undoubtedly improved its relationship with Apple. LG smartphones had carved out a notable slice of the US market, where it ranked No. 3 with a share of 10 per cent (compared with a 2 per cent share worldwide). It is easy to imagine that Apple would have regarded this confrontation on its home turf with displeasure. It is therefore no surprise that LG decided its loss-making toehold in smartphones was not worth maintaining if it jeopardised LG Innotek's ability to provide future components for the iPhone.

APPENDIX 5

CML'S ACTUATOR STRATEGY: AN OVERVIEW

2012	CML was unable to agree terms with a major smartphone manufacturer for the substitution of SMA OIS for VCM.
2013–14	CML was preoccupied with getting Hutchison Technology Inc. into production for the up-and-coming Chinese customers it had identified, including Huawei.
2015–16	Extensive discussions with Samsung resulted in a licence agreement in December 2015. However, in the wake of the Galaxy Note 7 battery fires of August 2016, the relationship evaporated and never regained its footing.
2017–19	CML achieved design wins for the P and Mate series from Huawei. Subsequently, the second tranche of American sanctions completely negated the company's handset projections and, by 2021, CML was forced to accept that Huawei was blighted.
2020–22	Faced with the plunge in Huawei's offtake, CML's thinking naturally turned to other ways of securing volume. A strategy of aligning itself with a strong and credible existing supplier to avoid the essential imbalance in the relationship between a leader in the smartphone business and a small company made sense.
2023	To break out of the confines of the business model it had constructed – and raise the value accruing to CML from each unit sold – the company decided to pursue a silicon strategy. This began to have an impact in 2025.

2024 Reinvigorated by its success in overcoming its chip supply issues, Huawei regained market share in China at the expense of Apple. Against this, geopolitics started to cast a longer shadow over CML's future relationship with Huawei.

2025 The contract engineering work CML had been engaged on since 2022 for VR/AR headsets and glasses gathered momentum. Additionally, CML's endeavours to diversify the market for SMA actuators began to show promise with active interest from the medical device industry.

ACKNOWLEDGEMENTS

My long association with Cambridge Mechatronics Limited as an investor inspired me to write this book. Directors, former directors and employees of the company gave me valuable assistance on the company and its technology and helped me to find my way round the smartphone industry. I particularly wish to thank Dr Tony Hooley, Dr Martin Knight, Andrew Osmant and Dominic Webber for their help and advice. Others beyond the company were kind enough to read portions of the text and tell me whether I was on the right path. I particularly wish to thank Ralph Cantellow, Andrew Doubt, Andrew Ellam and Gilbert Short and my fellow investors, Philip Bradley and Fiorangelo Salvatorelli. Kiana Palombo and the team at Whitefox guided me steadily and efficiently through the publishing process, for which I am most grateful, Any errors, oversimplifications, omissions or misjudgements are my responsibility and mine alone.

BOOK SOURCES AND BIBLIOGRAPHY

In writing this book I have relied heavily on online sources: corporate, media, industry-specific and research organisation websites and, particularly, the trove of information posted by the members of the techno-commentariat. All these sites are readily accessible to anyone who wishes to seek them out. As the text covers an industry that is relatively new and changes fast, few relevant books have been written. Those that I found helpful are listed below.

Ashton, James, *The Everything Blueprint: The Microchip Design that Changed the World*, Hodder & Stoughton, 2023

Capri, Alex, *Techno-Nationalism: How It's Reshaping Trade, Geopolitics and Society*, Wiley, 2024

Conway, Ed, *Material World: A Substantial Story of Our Past and Future,* W.H. Allen, 2023

Dou, Eva, *House of Huawei: The Secret History of China's Most Powerful Company,* Penguin Random House UK, 2025

Frier, Sarah, *No Filter: The Inside Story of Instagram,* Penguin Random House UK, 2020

McGee, Patrick, *Apple in China: The Capture of the World's Greatest Company,* Simon & Schuster, 2025

Miller, Chris, *Chip War: The Fight for the World's Most Critical Technology*, Simon & Schuster, 2023

Minshall, Tim, *Your Life Is Manufactured: How We Make Things, Why It Matters and How We Can Do It Better*, Faber & Faber, 2025

Wang, Dan, *Breakneck: China's Quest to Engineer the Future*, Allen Lane, 2025

IMAGE CREDITS

p. 21: Helimorph. Reproduced by permission of Cambridge Mechatronics Ltd (CML).

p. 29: 1Digital Sound Projector. Reproduced by permission of Cambridge Mechatronics Ltd (CML).

p. 32: Process Node Level Range. Reproduced by Tony Golding. Adapted from 'Semiconductor 101', Abhishek Singh, Medium, 2 July 2023.

p. 35: Stages of semiconductor production. Reproduced by permission of Bits&Chips.

p. 43: ASML EUV lithography system. Reproduced by permission of ASML.

p. 48: Yamaha YSP-1. Tony Golding.

p. 58: Voice Coil Motor in cross section. Courtesy of *Image Sensors World* (image-sensors-world.blogspot.com).

p. 65: Global smartphone operating system market share (2024–2025). Reproduced by permission of Counterpoint Research.

p. 66: Martin Cooper with his phone, Wikimedia Commons. Free use. Rico Shen, 5 June 2007.

p. 69: Premium smartphone sales trends to 2023. Reproduced by permission of Counterpoint Research.

p. 75: Global smartphone market shares, 2025. Reproduced by permission of Omdia.

p. 90: Semiconductor revenue forecast by product category. Reproduced by permission of Counterpoint Research.

p. 92: Snapdragon mobile processor. Reproduced by permission of Pocket-lint.

p. 97: Inside of an iPhone. Reproduced by permission of The Verge.

p. 103: Smartphone drop-test equipment. Reproduced by permission of Labtone Test Equipment Co.

p. 112: Camera module chain, 2010. Reproduced by permission of Cambridge Mechatronics Ltd (CML).

p. 122: Driver IC and match. Reproduced by permission of Cambridge Mechatronics Ltd (CML).

p. 125: Camera module lineup. Reproduced by permission of Samsung Electro-Mechanics Company (SEMCO).

p. 129: 'Computational photography'. Reproduced by permission of Vasily Zubarov.

p. 136: Camera module constituent parts. Reproduced by permission of Cambridge Mechatronics Ltd (CML).

p. 146: Emerging supply chains. Tony Golding.

p. 167: CML ownership. Reproduced by permission of Cambridge Mechatronics Ltd (CML).

p. 239: Periscope camera. Galaxy S20 Ultra, Samsung Mobile Press, 19 February 2020.

p. 241: Number of lenses per smartphone: historical trend. Reproduced by permission of Omdia.

p. 251: Top Smartphones by Camera Score, 2025, from DXOMARK (dxomark.com).

p. 252: Minutes of video watched by year, from 'Content Distribution: The Marketing Executive's Guide', Terakeet, 22 October 2021. Reproduced by permission of Terakeet.

p. 283: Apple versus Huawei smartphone market share in China. Reproduced by permission of Counterpoint Research.

p. 303: Apple Vision Pro, CNET, 7 June, 2023. Reprinted with permission. © 2026 Ziff Davis, LLC. All Rights Reserved.

p. 306: Meta's Ray-Ban Smart Glasses. Image by Prakhar Khanna, from ZDNET, 'Why Meta's Ray-Ban Smart Glasses are my favorite tech purchase this year', 13 May 2024.

p. 329: Image marking venture-capital funding completion. Reproduced by permission of Acuity Advisors.

INDEX

www.ingramcontent.com/pod-product-compliance
Lightning Source LLC
LaVergne TN
LVHW010711061126
851857LV00014B/1135
9781918191639